Simulating Nature

A Philosophical Study of Computer-Simulation Uncertainties and Their Role in Climate Science and Policy Advice

SECOND EDITION

Simulating Nature

A Philosophical Study of Computer-Simulation Uncertainties and Their Role in Climate Science and Policy Advice

SECOND EDITION

Arthur C. Petersen

CRC Press
Taylor & Francis Group
Boca Raton London New York

CRC Press is an imprint of the
Taylor & Francis Group, an **informa** business

A CHAPMAN & HALL BOOK

First edition published by Het Spinhuis Publishers, Apeldoorn, The Netherlands, 2006

CRC Press
Taylor & Francis Group
6000 Broken Sound Parkway NW, Suite 300
Boca Raton, FL 33487-2742

Library of Congress Cataloging-in-Publication Data

Petersen, Arthur C. (Arthur Caesar), 1970-
Simulating nature : a philosophical study of computer-simulation uncertainties and their role in climate science and policy advice / Arthur C. Petersen. -- 2nd ed.
        p. cm.
"A CRC title."
Includes bibliographical references and index.
ISBN 978-1-4665-0062-4 (alk. paper)
1. Climatology--Computer simulation. 2. Uncertainty (Information theory). 3. Climatic changes--Government policy. I. Title.

QC981.45.P48 2012
551.501'13--dc23   2011046557

Visit the Taylor & Francis Web site at
http://www.taylorandfrancis.com

and the CRC Press Web site at
http://www.crcpress.com

Contents

Section II   The Case of Simulating Climate Change

Preface to the Second Edition

Over the past few years, I have received numerous stimulating reactions to the first edition of this book. Especially within circles where uncertainties in climate simulations are intensively discussed, the hard-to-get paperback book—published by a small Dutch press with limited international exposure—with its striking orange cover seemed to have reached some 'cult' status. Just when it became clear that the book soon would be out of print, I was coincidentally approached by Rob Calver, editor for Chapman & Hall/CRC, who asked me to consider writing a book on modelling uncertainty in policy. I proposed instead to rework *Simulating Nature* into a second edition, which is how this edition came about.

Besides the introduction of corrections throughout the book, the inclusion of references to a number of recent publications by others and me on the topic of the book, the main change in this second edition is a thorough update and rewriting of sections that refer to the Intergovernmental Panel on Climate Change (IPCC) and its reports. Furthermore, I have lifted the anonymity of the country that caused the IPCC, for the first time, in its Summary for Policymakers in 2001, to give a quantitative estimate of the human contribution of the observed warming.

The first edition covered developments in the IPCC until 2006, just before the Fourth Assessment Report (AR4) was finalised in 2007. Where relevant, the latest report is now referred to. However, given the historic role of the IPCC's Third Assessment Report (TAR) of 2001 and the absence of fundamental changes in the reliability of climate simulations (my claim), the case study on the causal attribution of climate change in the TAR has retained its central place in the book. Through the Dutch government I had close involvement with the proceedings of IPCC since 2007 (among other things, I again attended the Working Group I (WG I) Plenary in Paris, January–February 2007, to approve the AR4 WG I Summary for Policymakers), and I co-led the Dutch assessment of IPCC AR4's statements on projected regional impacts of climate change in 2010 (after errors had been discovered), so this second edition benefits from the insights obtained over the past few years on the IPCC's assessment and communication of uncertainty. I hope this second edition is now even better positioned to contribute to an understanding of the working of the IPCC and to further improvements in its uncertainty assessment and communication during its Fifth Assessment Report (AR5), scheduled to appear in 2013–2014, and beyond.

Parts of the updated Chapter 7 have been published as a book chapter, 'Climate Simulation, Uncertainty, and Policy Advice: The Case of the IPCC', in G. Gramelsberger and J. Feichter (Eds.), *Climate Change and Policy: The Calculability of Climate Change and the Challenge of Uncertainty*, pp. 91–111 (© 2011

Springer), and parts of the updated Chapters 5 and 7 have been published as 'The Practice of Climate Simulation and Its Social and Political Context', *Netherlands Journal of Geosciences—Geologie en Mijnbouw* 87: 219–229 (2008).

Finally, I would like to thank Professor Leonard Smith, director of the Centre for the Analysis of Time Series (CATS), London School of Economics and Political Science (LSE), for encouraging my research in the philosophy of climate science, for fruitful discussions, and for providing an enabling environment during my stays at LSE as a visiting professor in CATS and the Grantham Research Institute on Climate Change and the Environment.

London, April 2012

Preface to the First Edition

The issue of uncertainty in science and policy has intrigued me for already more than a decade. After I had completed my MA thesis on consensus in climate science in the summer of 1995, I quickly discovered that the Dutch climate scientists with whom I did my atmospheric research were very open about the uncertainties in their science and were genuinely interested in my philosophical analyses. This stimulated me to ask my former thesis supervisor, Hans Radder, about the possibilities for postdoctoral research in philosophy of science. Hans was successful in obtaining university funding for a three-year project on the role of computer simulation in science and politics. Thus, after I had finished my dissertation in atmospheric science at Utrecht University, I returned to my alma mater, the VU University Amsterdam, as a postdoctoral research associate in April 1999. It was decided that the result of the project should be a second dissertation, and that I should also enrol in the Netherlands Graduate School of Science, Technology and Modern Culture (WTMC). To some it may seem odd that a postdoctoral researcher returns to graduate school, but I found this period to be intellectually one of the most stimulating and fruitful periods of my career, and I still retain many fond memories of the people I met at the workshops, the summer schools and the winter school that I attended. I enjoyed being part of the faculty of philosophy, and I would like to thank both my dissertation supervisors, Professors Hans Radder and Peter Kirschenmann, for guiding me through this project, for their incisively critical comments that always led to significant improvements in my thinking and the text, and for their confidence that I would—sometime—complete this project. Simply put, without them this book would not exist. Several other people provided helpful material and comments that stimulated my work. I would like to mention in particular Marcel Boumans, Silvio Funtowicz, Stephan Hartmann, Peter Janssen, Chunglin Kwa, Harro Maas, Mary Morgan, Jerome Ravetz, James Risbey, Sergio Sismondo, Marjolein van Asselt, and Jeroen van der Sluijs.

When I was about halfway through the project, I obtained a position as senior scientist in uncertainty assessment and communication at the Netherlands Environmental Assessment Agency, and I left the VU University Amsterdam on 1 January 2002. The new job turned out to be beneficial for the project. Although it took nearly another five years to finish this study, I feel that the ideas that are now laid out in it are much more mature than what I could have offered after three years of research as an outsider. The two directors of the Netherlands Environmental Assessment Agency, Klaas van Egmond and Fred Langeweg, have given strong support to the different research activities that I have undertaken within the agency. There are many other agency colleagues whom I should mention here. However, I will limit myself to thanking four colleagues who contributed most to this study: Arthur Beusen, Peter Janssen, Johan Melse, and Anton van der Giessen.

In this study, aside from analysing the role of simulation in natural science and public policy in general, I focus on climate science and policy in particular. This choice is obviously related to my own disciplinary background, but also to the high societal and political importance of appropriately assessing and communicating uncertainties in climate simulation. My MA thesis already addressed the Intergovernmental Panel on Climate Change (IPCC) and the controversial nature of its Summaries for Policymakers. How can one formulate a 'consensus' about what we know about climate change if climate scientists in their daily scientific practice seem to disagree on so many points, particularly with respect to what constitutes a 'good' climate model? For my studies of climate science and policy, I am particularly indebted to the following people, who were interviewees or advisers: Fons Baede, Henk Dijkstra, David Griggs, John Mitchell, James Risbey, Cor Schuurmans, Rob Swart, Paul van der Linden, Aad van Ulden, Koos Verbeek, and Hans von Storch.

Aside from my dissertation supervisors, many people gave comments on different portions of the manuscript in its various stages. I would like to thank in particular Ben Bakker, Henk de Regt, Maarten Kleinhans, Martin Krayer von Krauss, Chunglin Kwa, Andrea Scharnhorst, Frits Schipper, Paul Wouters, and again Sergio Sismondo and Marjolein van Asselt.

Finally, I would like to acknowledge copyrighted material and additional financial support. Parts of Chapters 1 and 4 have been published as a book chapter titled 'Simulation Uncertainty and the Challenge of Postnormal Science', in J. Lenhard et al. (Eds.), *Simulation: Pragmatic Constructions of Reality—Sociology of the Sciences*, vol. 25, pp. 173–185 (© 2006 Springer). In addition to support from the VU University Amsterdam and the Netherlands Environmental Assessment Agency, the Vereniging voor christelijk wetenschappelijk onderwijs, the American Geophysical Union, and the Dutch National Research Programme on Global Air Pollution and Climate Change made it possible to attend several conferences and IPCC meetings. And the Netherlands Graduate School of Science, Technology and Modern Culture (WTMC) financially contributed to the publication of this study.

Bilthoven, October 2006

About the Author

Arthur C. Petersen (1970) obtained PhD degrees in atmospheric physics and chemistry (Utrecht University, 1999) and in philosophy of science (VU University Amsterdam, 2006). He now works as chief scientist at the PBL Netherlands Environmental Assessment Agency and as special professor of science and environmental public policy at the VU University Amsterdam. He is visiting professor at the London School of Economics and Political Science and research affiliate at the Massachusetts Institute of Technology.

After his dissertation in philosophy of science, on uncertainty in computer simulations of climate change, Arthur continued researching topics in methodology and modelling. He published on methodological aspects of the Intergovernmental Panel on Climate Change, the statistics of climate change in the past and the value ladenness of model assumptions. He also performed social-scientific research into (methods for) responsibly dealing with uncertainty and has published on 'postnormal science', stakeholder participation in scientific assessments, dealing with value plurality, adaptive policymaking and dealing with uncertain technological risks.

Besides his professional jobs, Arthur has been active within Pugwash (or Pugwash Conferences on Science and World Affairs in full), an organisation that brings together, from around the world, influential scholars and public figures concerned with reducing the danger of armed conflict and seeking cooperative solutions for global problems such as those related to poverty alleviation and protection of the environment.

List of Abbreviations and Acronyms

1-D: one dimensional
3-D: three dimensional
4-D: four dimensional
AGCM: Atmospheric General Circulation Model
AOGCM: Atmosphere–Ocean General Circulation Model
AR4: Fourth Assessment Report
AR5: Fifth Assessment Report
CA: cellular automata
COMEST: World Commission on the Ethics of Scientific Knowledge and Technology
CRM: cloud-resolving model
CSM: climate-system model
DNS: direct numerical simulation
EEC: European Economic Community
FAO: Food and Agriculture Organisation
FORTRAN: Formula Translator/Translation
GCM: general circulation model
GDP: gross domestic product
GFDL: Geophysical Fluid Dynamics Laboratory
IDL: Interactive Data Language
IGBP: International Geosphere–Biosphere Programme
IMAU: Institute for Marine and Atmospheric Research Utrecht
IPCC: Intergovernmental Panel on Climate Change
LES: large-eddy simulation
MM: mesoscale model
MNP: Milieu-en Natuurplanbureau (Netherlands Environmental Assessment Agency until 2008)
NCAR: National Center for Atmospheric Research
NRC: U.S. National Research Council
NUSAP: Numeral Assessment Spread Assessment Pedigree
OS: operating system
PBL: Planbureau voor de Leefomgeving (Netherlands Environmental Assessment Agency since 2008)
PBL: planetary boundary layer
RIVM: Rijksinstituut voor Volksgezondheid en Milieu (National Institute for Public Health and the Environment)
SAR: Second Assessment Report
SBSTA: Subsidiary Body for Scientific and Technological Advice
SPM: Summary for Policymakers
SRES: Special Report on Emission Scenarios

TAR: Third Assessment Report
TARGETS: Tool to Assess Regional and Global Environmental and Health
          Targets for Sustainability
TS: Technical Summary
TSU: Technical Support Unit
UN: United Nations
UNEP: United Nations Environment Programme
UNESCO: United Nations Educational, Scientific and Cultural Organisation
UNFCCC: United Nations Framework Convention on Climate Change
UNIX: Uniplexed Information and Computing System
WG: Working Group

1

Introduction

1.1 Background

On January 20, 1999, Dr. Hans de Kwaadsteniet, a senior statistician at the Netherlands National Institute for Public Health and the Environment (Rijksinstituut voor Volksgezondheid en Milieu, RIVM), made news in the Netherlands. After years of trying to convince his superiors that the environmental assessment branch (Netherlands Environmental Assessment Agency)* of the institute leaned too much toward computer simulation at the expense of measurements, he went public with this criticism by publishing an article on the op-ed page of the national newspaper, *Trouw* (de Kwaadsteniet 1999). His article was supplemented with an interview that resulted in the headline 'Environmental Institute Lies and Deceives' on the newspaper's front page. His specific claim was that the institute suggested too high an accuracy of the environmental figures published yearly in its *State of the Environment* report. According to him, too many model results that had not been rigorously compared with observational data were included—mostly because of the lack of sufficiently detailed data with which to do the necessary comparisons. He pointed out that living in an 'imaginary world' was dangerous. He thought that if the institute spent more time and energy on testing and developing computer-simulation models in a way that made greater use of existing and newly performed observations, it would become more careful in the way it presented its results to policymakers. De Kwaadsteniet identified the deceptive speed, clarity, and internal consistency of the computer-simulation approach as the main causes of the claimed bias toward computer simulation at the RIVM's environmental assessment branch.

---

* The Netherlands Environmental Assessment Agency is a governmental body that by statute provides the Dutch government and parliament—and the European Commission, European Parliament, and U.N. organisations—with scientific advice on environmental, sustainability, and spatial planning problems. Its acronym has changed over the past decade from RIVM (Rijksinstituut voor Volksgezondheid en Milieu, of which the agency's function was a part), through MNP (Milieu-en Natuurplanbureau, since 2003), to PBL (Planbureau voor de Leefomgeving, the Dutch name since May 2008, when it merged with the assessment agency RPB, to include assessments of spatial planning issues). The English name has remained unchanged since 2003.

   The institute responded immediately to the publication by suspending de Kwaadsteniet from his job and stating in an official reaction that a significant fraction of its environmental research budget was spent on observations, that the main policy-relevant conclusions in the institute's reports were robust in the light of uncertainties, and that the uncertainties were not deliberately left out of the *State of the Environment* reports. The institute promised to publish material concerning the uncertainties in future editions. In a later reaction, the RIVM's director of environment, Professor Klaas van Egmond,* argued that simulation models had to be be viewed as 'condensed knowledge', and that they were indispensable in environmental assessment since without them it would be impossible to determine cause–effect relationships between sources and effects of pollution (van Egmond 1999). Thus, models gave meaning to measurement results, he added. And they were needed in environmental policymaking. Furthermore, he observed that policymakers were often confronted with incomplete knowledge, and that the institute regarded it as its task to report on the current state of affairs in the environment, including the uncertainties involved. As an example, he stated that it would take many years before climate research reached the 'ultimate truth' about what was happening to the climate. However, based on currently available knowledge and its uncertainties, politicians have to decide on whether to take measures now. Finally, the director added that, for policymakers, the most important conclusions contained in the summaries of the *State of the Environment* reports were carefully crafted, taking all relevant uncertainties into account.
   Soon after the publication of de Kwaadsteniet's article, an intense and long-lasting media debate ensued in the Netherlands.† The affair reached the floor of the Dutch Parliament within a matter of days. Facing the Parliament, the minister of the environment, Jan Pronk, defended the integrity of the institute. In return for an agreement that the institute would organise more regular external reviews of its environmental assessment activities and improve its communication of uncertainty, the minister granted the institute additional funding for its monitoring activities.

                                         ***

On June 7, 1999, I publicly defended my doctoral dissertation "Convection and Chemistry in the Atmospheric Boundary Layer" (Petersen 1999b). In this dissertation, the main body of which consisted of three journal articles based on computer simulation (Petersen et al. 1999; Petersen and Holtslag 1999; Petersen 2000a), I argued that one of the uncertainties in regional and global computer models of air quality was significantly smaller than was previously thought. Formerly, it was not known whether the influence of

---

* Subsequently, Klaas van Egmond was director of the Netherlands Environmental Assessment Agency (MNP) until the end of 2007.
† See van Asselt (2000) and van der Sluijs (2002) for more information about this debate.

turbulence on chemical reactions in the atmospheric boundary layer could be neglected. I, together with my colleagues from the Institute for Marine and Atmospheric Research Utrecht (IMAU), using a hierarchy of computer models, had shown that this neglect was allowable. One of the opponents, Professor Frans Nieuwstadt of Delft University of Technology, sternly questioned me about the reliability of my research results until he was satisfied with my final answer that I was confident about my research results only within a factor of two.* His main problem with the work was that only simulation models of different complexity had been compared with each other, and no comparison had been made with experimental or observational data. My contention was that the most complex simulations that I had done using the national supercomputer of the Netherlands were more reliable for answering my research questions than were any of the sparse experimental or observational results reported in the literature. This was judged by Professor Nieuwstadt to be a 'medieval position'. I disagreed since the large-eddy simulation (LES) model that I had used had been rigorously compared with experimental and observational data.† The only thing I had done, I claimed, was to apply this model to a somewhat different problem, which was extremely difficult to approach experimentally or observationally. After this minor public controversy, the episode ended well since the doctorate was awarded by the committee without any objections.‡

1.2 Framing of the Problem

The two episodes are by no means isolated examples of controversies concerning the uncertainties in computer simulation. De Kwaadsteniet questioned the reliability of scientific simulation results for use in politics. Nieuwstadt was concerned with their use for purely scientific purposes. All around the world, both within and without science, controversies like the two described here have regularly surfaced. Since the introduction of computer simulation in science in the 1940s, for instance, scientists have held 'debates that continued decades later over the legitimacy of according doctorates to students who had "only" simulated experiments' (Galison 1996: 155). The use of computer-simulation results in policymaking is also regularly questioned in political circles. For instance, in 1995, U.S. politicians—to

---

* This accuracy was high enough for drawing the conclusions that I wanted to draw.
† On this, Professor Nieuwstadt had to agree. Although I did not bring this into the discussion, I knew that Nieuwstadt was well aware of this fact since the LES model was his.
‡ For those unfamiliar with the Dutch university system, as in many other countries, PhD students are admitted to defend their dissertation only after an assessment committee has approved it. Nieuwstadt, who was an external member of this committee, had already approved it.

be precise, the Republican majority of the House of Representatives—even proposed legislation (which did not get through Congress in the end) to base environmental regulation on observations only and to officially sideline any model result in the policymaking process.

The practice of scientific simulation and the role of simulation in science and policy give rise to a wide range of philosophical questions. Scientific computer simulation is portrayed by some philosophers of science as a new method of doing science, besides theorising and experimentation (e.g., Rohrlich 1991; Humphreys 1994; Keller 2003; Winsberg 2010). Science studies generally seem to support this conclusion from a historical or sociological perspective (e.g., Galison 1996; Dowling 1999). Two major reasons are typically given for why simulation should be considered qualitatively different. First, it is claimed that simulations make it possible to 'experiment' with theories in a new way. For instance, Deborah Dowling, based on interviews with scientists who practice simulation—let us call them 'simulationists'—describes the practice of simulation as follows:

> By combining an analytical grasp of a mathematical model with the ability to temporarily 'black-box' the digital manipulation of that model, the technique of simulation allows creative and experimental 'playing around' with an otherwise impenetrable set of equations, to notice its quirks or unexpected outcomes. The results of a large and complex set of computations are thus presented in a way that brings the skills of an observant experimenter to the development of mathematical theory. … In their everyday interactions with the computer, and in their choice of language in varied narrative contexts, scientists strategically manage simulation's flexible position with respect to 'theory' and 'experiment.' (Dowling 1999: 271)

Second, simulation enables us to extend our limited mathematical abilities so that we can now perform calculations that were hitherto unfeasible. Thus, we can both construct new theories using computer simulation and calculate the consequences of old theories.[*] An example of the former category is the application of cellular automata in biology (Rohrlich 1991; Keller 2003). The latter category is exemplified by the study of turbulent flows based on the well-established nonlinear equations of fluid dynamics, the Navier–Stokes equations. Although there is no fundamental difference between the computability of problems before and after computers became available (an argument against putting simulation in a philosophically distinct category; cf. Frigg and Reiss 2009), through the introduction of the computer an actual barrier in scientific practice to the large-scale use of numerical mathematics—that is, the

---

[*] Both 'new' aspects of scientific research are made possible by the high speed with which computers perform calculations. And both aspects are interrelated: We can derive new theories by 'playing' with variants of old theories, and the calculation of results for the new theories is often unfeasible without using the computer.

limited speed with which humans, even if aided by mechanical machines, can do calculations—was removed. I agree with Paul Humphreys:

> While much of philosophy of science is concerned with what can be done in principle, for the issue of scientific progress what is important is what can be done in practice at any given stage of scientific development. (1991: 499)

I make my own philosophical commitments more explicit in Section 1.4.

Philosophical analysis can contribute to a deeper understanding of the controversies concerning uncertainty in simulation. In Section I of this study, I address the following three general questions:

1. What specific types of uncertainty are associated with scientific simulation?
2. What are the differences and similarities between simulation uncertainty and experimental uncertainty?
3. What are appropriate ways to assess and communicate scientific simulation uncertainties in science-for-policy?

To answer these questions, empirical results from science studies and political science are used to inform the philosophical analysis.

The scope of Section I includes all fields of science in which nature (both physical and biological) is simulated, while the main examples are drawn from the earth sciences. Computer simulation as a scientific approach is not limited to the natural sciences, however: Simulation is gaining ever more prominence, for example, in psychology, sociology, political science, and economics.* The simulation of human behaviour (individual or collective) gives rise to additional questions related to the capacity of humans to reflect. This reflexive capacity adds a qualitatively different source of uncertainty and unpredictability to computer simulation as compared with simulations of nature. However, questions that focus on the specific uncertainties in simulating human behaviour fall outside the scope of the present study. Still, since the main examples in this book are instances of complex systems, part of the analysis presented also applies to many social science simulations.

With respect to areas of policymaking in which simulation results are used, the scope of Section I is broad. It includes topics as diverse as nuclear weapons policies (using, e.g., physics simulations); environmental pollution policies (using, e.g., toxicology simulations); biodiversity policies (using, e.g., ecological simulations); drug policies (using, e.g., biomedical simulations), and so on.

---

* The recent rise in the amount of work on simulation in these fields may be partly related to the wide applicability of the concept of 'complex systems' (see Casti 1997, who provides a popularised account of the use of simulation in the natural and social sciences to study complex systems). Many simulations in both the natural and social sciences share system-theoretical concepts.

In Section II, one policy arena in which these questions figure most prominently is highlighted, that of anthropogenic climate change. Are humans
currently changing the climate? Ask policymakers, and they will probably
reply that climate scientists, in the context of the Intergovernmental Panel on
Climate Change (IPCC), have indeed drawn a positive conclusion. How do
scientists reach such conclusions, and why do policymakers trust them? An
adequate answer to these last two questions requires both sociological and
philosophical research on the role of models in science and on the translation
of model results into a political context (see also Petersen 2000b).

The dominance of one specific way to frame the climate-change problem, that of using global climate models to project future climate change for
different input scenarios (e.g., of greenhouse gas emissions) and different
model assumptions, as done by the IPCC, has been criticised in the social-
scientific literature (Jasanoff and Wynne 1998). Especially in the early years
of the IPCC, the often publicly voiced criticism of climate models by 'greenhouse sceptics' was said not to have been adequately dealt with or clearly
reflected in the summaries for policymakers. For the IPCC, this was a reason
to try to improve the review procedures. An evaluation of whether the IPCC
has succeeded is of crucial importance for the legitimacy of climate policies.
The method adopted in this study for evaluating the IPCC is to apply the
philosophical insights concerning simulation uncertainty gained in Section
I to climate simulation. Thus, the aim is not a full evaluation of the IPCC. The
following specific questions are addressed in Section II:

4. What specific types of uncertainty are associated with the simulation-based attribution of climate change to human influences?

5. Have these uncertainties been appropriately assessed and communicated by the IPCC?

As will become clear, the issue of climate-simulation uncertainty played a
pivotal role in constructing the evolving conclusion on the human influence
on recent global warming by IPCC Working Group I, in its second, third, and
fourth assessment reports of 1996, 2001, and 2007a. In this book, I zoom in on
the 2001 report for a case study since in that report the transition was made
to a probabilistic statement about the human influence on historic warming,
based on data, climate-simulation models, and expert judgement.

1.3 Defining Computer Simulation and
Positioning It in Science

Before I turn to explaining the philosophical approach of this study, I
introduce a definition of *computer simulation* and briefly discuss historical,

sociological, and philosophical work on the question of the role of computer simulation in science.

The term *computer simulation* does not seem to have a sharply defined meaning in scientific practice. Furthermore, analysts of scientific practice do not stick to one definition. *Computer simulation* is often used interchangeably with *numerical experiment* (e.g., Naylor 1966; Galison 1996; Dowling 1999; Winsberg 2003). Sometimes, however, the word *experiment* is omitted, and broader definitions are given, such as 'any computer-implemented method for exploring the properties of mathematical models where analytic methods are unavailable' (e.g., Humphreys 1991: 501). Also, narrower definitions have been introduced, which do not simply equate simulations with numerical experiments but restrict the term to a subclass of numerical experiments, usually those that also satisfy the following characterisation:

> A simulation imitates one process by another process. (Hartmann 1996: 83)*

I follow this definition with two caveats.

First, one of the common parlance associations with the verb *imitate*, namely, that imitation is not something virtuous, is not implied. By being linked to *imitation*, the term *simulation* can indeed easily share the negative connotations of imitation. For example, Nancy Cartwright, based on an entry in the *Oxford English Dictionary*, defines a *simulacrum* (an old-fashioned word for 'simulating thing') as

> something having *merely* the form or appearance of a certain thing, without possessing its substance or proper qualities. (Cartwright 1983: 152–153; emphasis added)

Of course, in scientific simulation one has to be aware of the fact that the imitation may be 'fooling' the researcher. Many are aware of this fact, as can be witnessed from the controversies about the reliability of computer simulation. Still, it must be kept in mind that the word *simulation* also became used as a positive term in science after World War II (Keller 2003: 198–199).

Second, it is not implied that both processes are isomorphic. The imitation relation only entails that the outcomes of a simulation process mimic

---

* In Hartmann's description, the term *simulation* is not intended to be limited to computer simulation. Here, we are only interested in computer simulation, however. What is entailed by the 'imitation relation' is discussed in the main text that follows.

relevant features of a real-world process.* A simulation is a *representation of* a real-world process (Morgan 2003: 229).†

The computer simulations considered in this study all involve a mathematical model that is implemented on a computer and imitates a process in nature. It is assumed that phenomena in the real world are under study and that the simulation consists of a numerical process that aims at imitating these realised or potential (e.g., counterfactual or future) phenomena.‡ The mathematical model together with its conceptual interpretation constitute a theoretical model.§ A computer-simulation model can thus be regarded as a theoretical model materialised in a computer. The theoretical model consists of numerical mathematical equations and logical operations, on the one hand, and the conceptual interpretation of the mathematics on the other.

Now that we have clarified some definitions, let us take a look at the history of computer simulation. This history is obviously directly tied to the history of the computer, and both are strongly coupled to military history. For instance, the successful test of the first hydrogen bomb in 1952, which would have been impossible to construct without the use of computer simulation (see Galison 1996), did give much credibility to the new scientific practice of computer simulation within the physics community. Moreover, the development of numerical weather prediction serves as a unique example of a military-funded coordinated effort leading to a significant scientific advance, with direct civilian benefits (Harper 2003).

Since World War II, simulation approaches in science have emerged and expanded—not in isolation, but often in strong contact with experimental and observational fields in the natural sciences and aided by developments in mathematics and computer science. Scientists use tools to gain knowledge of the world. The computer is one of these tools. Indeed, the introduction of the computer has had a significant influence on the development of science

---

* It is assumed that there is a real world and that the relationship between a simulation and this real world—no matter how this relationship is conceived—is of interest to both scientists and analysts of scientific practice. The definition does not imply that the possible real-world processes to be imitated should be realised, or even practically realisable, in nature. When simulation is used to study the workings of processes under artificial conditions, there is still an assumed relationship with the possible behaviour of the real world under those conditions, even if the conditions are not realisable in practice.

† Morgan (2003) claims that some simulations are also *representative of* and *representative for* real-world processes. This position is investigated in Chapter 2.

‡ It is not implied here that the detailed specification of the process in the computer should realistically represent all the details of the real process.

§ Bailer-Jones and Hartmann (1999) distinguish between three different kinds of models that play a role in science: *scale models, analogical models* (based on certain relevant similarities), and *theoretical models*. Usually, scale models and analogical models are *material*. Theoretical models are of an *abstract-theoretical* nature. However, other subdivisions can be made, and above all, such distinctions between kinds of models should not be interpreted as absolute. For example, a particular model can be an analogical and a theoretical model at the same time. This is true, for instance, for the 19th century billiard-ball model of gases (kinetic theory), which was extensively discussed by Hesse (1963).

in the 20th century.* Actually, in scientific practice, computers are deployed for many different functions. For instance, they are used to control experiments, to store and visualise data, to write and typeset scientific texts, to communicate with other scientists through e-mail and the World Wide Web, as well as to perform 'simulations'.

In the short history of scientific simulation, the list of problems that have been addressed using simulation has undergone significant expansion. Evelyn Fox Keller has suggested three stages in this history (Keller 2003: 202):

(1) the use of the computer to extract solutions from prescribed but mathematically intractable sets of equations by means of either conventional or novel methods of numerical analysis;

(2) the use of the computer to follow the dynamics of systems of idealized particles ... in order to identify the salient features required for physically realistic approximations (or models);

(3) the construction of models ... of phenomena for which no general theory exists and for which only rudimentary indications of the underlying dynamics of interaction are available.

Keller also offers her historical hypothesis as a typology of simulation. The three types of simulation practice can indeed all be discerned in present-day science.

Some examples may serve to illustrate the different types of simulation. Galison (1996) and Keller (2003: 203–204) describe the development of the technique of 'Monte Carlo simulation' in the early days of simulation—coinciding with the advent of the first electronic computers in the 1940s. This mathematical technique was originally developed for numerically integrating nonlinear equations describing nuclear detonations. It belongs to the first type of scientific simulation.[†] Later, in the 1950s, the time evolution of dynamical systems became the subject of scientific simulation. Keller gives molecular dynamics as an example of this second type of scientific simulation. Another example is numerical weather prediction. In the 1950s, simulations were developed to predict the weather a few days in advance. The 'idealised particles' in the early numerical weather prediction models were a few hundred fluid particles—each hundreds of kilometres in horizontal size—together making up the piece of atmosphere that was modelled (see, e.g., Nebeker 1995). Finally, the third type of simulation is closely tied by Keller to cellular automata and artificial life. She observes that '[d]espite initial hopes in the value of CA [cellular automata] modelling in promoting better theory—in particular, a better understanding of biological principles—Artificial Life studies have made little impression on practicing biologists' (Keller 2003: 213). One must add here that Keller's third type of

---

[*] This should remind us of the fact that scientific practices, including the specific elements that play key roles in them, remain in flux and change over time.

[†] A Monte Carlo simulation—although it may be used to perform calculations on the dynamics of systems of particles—differs from the second type of simulation since it does not keep track of a set of particles interacting with each other; that is, it does not explicitly include a system of particles in its imitation of the process studied.

simulation should not be limited to the technique of cellular automata. In most scientific fields, ranging from high-energy physics to ecology and bio-medicine, simulations are performed that are not rigorously built on theory (see, e.g., Dowling 1999).[*]

1.4 Philosophical Approach

This study presents the philosophical results of my practical and intellec-tual journey through several disciplines and institutions.[†] Although the primary intention is a contribution to the philosophy of science, the inter-disciplinary nature of the endeavour and of the present text will become apparent to the reader.

My aim is to offer a philosophical account of scientific computer simula-tion that is theoretical, normative, and reflexive (Radder 1996: 169–187). Let me first discuss the *theoretical* dimension of my approach. This is what makes a philosophical study theoretical, in the words of Hans Radder (1996: 170):

> Theoretical philosophy of science and technology, as I see it, endeavors to expose and examine structural features that explain or make sense of nonlocal patterns in the practices, processes and products of science and technology.

Thus, the conceptualisation of scientific simulation practice that I present in Chapter 2 is not meant to be a 'straightforward description of empirical pat-terns' (Radder 1996: 170) but rather a characterisation of some structural fea-tures of simulation that facilitate the derivation of epistemologically distinct types of uncertainty. I make use of my own experience in scientific simula-tion, as well as of social studies of scientific simulation practice, to inform the philosophical debate. I do not take a strong stand in several philosophi-cal debates (e.g., realism vs. instrumentalism; models vs. theories). For the purpose of what I set out to do, it suffices to identify those general debates.

---

[*] The director of the U.S. Department of Energy's Office of Science, Raymond Orbach, for instance, said in *Business Week*: 'I'm a theoretical physicist, and there are some problems for which there aren't any theories. You can only understand that science through simula-tions' (Port and Tashiro 2004; see http://www.businessweek.com/magazine/content/04_23/b3886002.htm).

[†] These include university departments (Vrije Universiteit Amsterdam: theoretical phys-ics, philosophy; Utrecht University: marine and atmospheric research; and again Vrije Universiteit: philosophy and—most recently—environmental studies), national graduate schools (atmospheric and marine research; science, technology and modern culture), a U.N. body (IPCC), a governmental agency (Netherlands Environmental Assessment Agency), and disciplinary fields (theoretical nuclear physics; philosophy of science; atmospheric physics and chemistry; social studies of science; political science; environmental science; uncertainty analysis; and sustainability assessment).

I consider these debates as partly embodied in scientific practice. Scientists hold philosophical positions, and these influence their epistemic values. Thus, I do provide entries into these debates for they are relevant to how the uncertainties of simulation results are assessed in practice.

The structural features of scientific simulation practice that are pivotal for my study are the methodological rules that scientists aim at or claim to follow in developing and evaluating simulations, on the one hand, and the epistemic value commitments of these scientists on the other hand. Following Laudan (1984), I assume that there is a fundamental plurality of factual claims, methodologies, and epistemic values in scientific practice. Thus dissensus in scientific practice may reside at three levels:

- factual level (factual claims);
- methodological level (methods of development and evaluation);
- axiological level (aims and goals of scientific practice).

The elements of scientific practice corresponding to these three levels—that is, facts, methods, and aims—form a 'triadic network of justification' (Laudan 1984: 63). This means that changes in one of the three levels can be justified by making reference to any of the other two levels. For establishing the reliability of a theory, reference can be made, for instance, to the methodological level (claims are then submitted to methodological 'tests'). If there is no consensus on which tests should be applied, this methodological disagreement may subsequently be related to different opinions about the aims and goals of simulation practice.* For the study of any given process by means of simulation, a multitude of simulation models can legitimately be used, even though the extent to which plurality is realised will vary depending on the context. It may be perfectly rational for simulationists to use different models, provided that they are willing to submit their models to each others' critical scrutiny. The model of rationality assumed here is that of Harold Brown (1988: 183–196), which he contrasts with a 'classical model of rationality' based on a notion of universally applicable rules.† In discussions in which scientists critically assess each others' claims and models, they may—but need not—arrive at explicit consensus. For scientists to be able to disagree rationally, however, they need to share a common practice that features elements of explicit and implicit consensus (Brown 1988: 209–210; Petersen 1995).

---

* Laudan argues that besides these two examples of justification processes (which are part of the standard 'hierarchical model of justification' in the philosophy of science), three other justification processes play a role in scientific practice (Laudan 1984: 63): (1) methods are constrained by factual claims; (2) methods must exhibit the realisability of aims (if aims cannot be realised, this counts against the aims); and (3) factual claims must harmonise with the aims. None of the levels is more fundamental than any other.

† Brown (1988: 193) states that 'on the model I am proposing, the predicate "rational" characterizes an individual's decisions and beliefs, it does not characterize propositions and it does not characterize communities'.

The notion of 'judgement' is central to Brown's model of rationality, and this entails 'that our ability to act as rational agents is limited by our expertise' (Brown 1988: 185). The capacity of making 'good' judgements is formed by scientific training. The writings of Michael Polanyi provide support for Brown's focus on judgement. Science is an art 'which cannot be specified in detail [and] cannot be transmitted by prescription, since no prescription for it exists; [i]t can be passed on only by example from master to apprentice' (Polanyi 1962: 53). Thus the rationality of scientists is bound to the different communities of scientists* and depends on their skills.† As Polanyi (1962: 60) put it:

> It is by his assimilation of the framework of science that the scientist makes sense of experience. This making sense of experience is a skilful act which impresses the personal participation of the scientist on the resultant knowledge.

Belonging to a scientific practice is a 'form of life'. Consequently, no conscious rational decisions need to be made by a scientist to deal with many of the elements of his or her scientific practice. And in rational discussions about scientific claims, a large part of the body of knowledge on which these claims are built remain hidden from view.

Having briefly introduced some of my philosophical commitments with regard to the epistemological dimension of scientific practice, I now turn to the social and political dimensions. Of course, besides epistemic values and aims, scientific practice features nonepistemic values and aims. These values and aims may interact with the epistemic values. For instance, if a scientist is asked to produce a model that is to be used in a civil engineering project, the scientist may be mainly interested in a model that 'works' and produces accurate results but need not contain a realistic description of the underlying processes. The same scientist may prefer to use a more realistic model for more fundamental scientific research. For this study into the role of simulation in science and policy, it can thus be expected that the context for which a model is made influences the epistemic values and hence the methodologies employed by the scientists who develop and evaluate scientific simulations. However, in the study of scientific practice it is not easy to distinguish between epistemic and nonepistemic aspects. As Joseph Rouse writes,

> [W]e cannot readily separate the epistemological and political dimensions of the sciences: the very practices that account for the growth of scientific knowledge must also be understood in political terms as power relations that traverse the sciences themselves and that have a powerful impact on our other practices and institutions and ultimately upon our understanding of ourselves. (1987: xi)

---

* Here, again, plurality enters. Science consists of many different, but interconnected, practices.
† See also Ravetz (1996: 75–108).

It is important to reflect on the political dimensions of scientific simulation practice, especially when we look at simulation results that are used in policymaking.

This leads us naturally into a discussion of the *normative* dimension of this study. The task of philosophical studies of science is generally taken to be

> to formulate criteria that make it possible to analyze, assess, and possibly criticize scientific methods and scientific knowledge from a point of view that is either outside science or at least more comprehensive than science. (Radder 1996: 175)

Such criteria should make it possible to determine what is 'good' or 'justified' science. In the history of the philosophy of science, the proposed criteria—as well as their purpose—have frequently changed. In Chapter 3, I develop an argument for a methodological and social interpretation of the notion of 'reliability' of simulation, which should enable scientists and others to assess the reliability of claims derived from simulation. My normative concerns have a wider scope than science alone, as can be seen in the treatment of my main research questions 3 and 5 (see Section 1.2). Answering these questions demands the formulation of criteria for 'appropriately' assessing and communicating uncertainties in science-for-policy.

Finally, the *reflexive* dimension of my philosophical approach relates to 'the fact that the philosophical accounts of the practice of technoscience are themselves part of that very practice' (Radder 1996: 185). Since the second part of this book highlights uncertainties about whether humans cause climate change, my study can be used by climate 'sceptics' as backing for their argument that we should not continue to implement climate policies since, so they claim, the science is 'too uncertain'. However, at the same time, it can be used to argue that since there is some simulation-based evidence of human-caused climate change, we should be precautious and make sure that we avoid 'too much climate change', even though uncertainty remains. Without disclosing my own political preference, I hope that this book may contribute to a more sophisticated public discussion of simulation uncertainties surrounding climate change. Currently, we have climate sceptics magnifying uncertainties and climate activists downplaying these.

To conclude this brief exposition of my philosophical approach, let me also be *self*-reflexive: I have been doing scientific simulation within a specialised field for four years; I have participated as a Dutch government delegate in meetings of the IPCC[*]; I have been closely involved in the development of a 'guidance' for assessing and communicating uncertainty in science-for-policy,

---

[*] I was admitted to the Dutch government delegation in 2001 as a 'philosophical observer'. Although I did participate in the discussion on several statements on climate-change observations, I did not participate in the discussions on (simulation-based) climate-change attribution analysed in Chapter 7. In 2007, after the first edition of this book had been published, I became a full member of the delegation.

and I am still responsible for the implementation of that guidance within the Netherlands Environmental Assessment Agency. I am thus situated in the midst of the practices that I seek to analyse. This situatedness has advantages and disadvantages. The main advantage is an intimate knowledge of these different practices. The main disadvantage is that my own perspective may remain too close to these practices at the risk of having missed important issues with respect to the role of simulation in science and policy.

1.5 Brief Outline of This Study

Section I addresses the scientific practice of simulation, the uncertainties involved in simulation, and the role of simulation and its uncertainties in policy advice. Chapter 2 analyses the practice of scientific simulation as a 'laboratory' and examines the four main elements of this practice along with some philosophical issues that touch on these elements. Furthermore, the plurality of methodologies and values in simulation practice is discussed. Finally, the practice of simulation is compared with the practice of experimentation. Chapter 3 presents a typology of uncertainty in simulation and discusses the various dimensions of this typology. Subsequently, the uncertainties in simulation practice are compared with uncertainties in experimentation practice. Chapter 4 addresses the use of scientific simulation in policymaking. After treating general issues related to the science–policy interface and the role of simulation, including the present condition of 'postnormal science', a new methodology for assessing and communicating uncertainty in science-for-policy, which was developed at the Netherlands Environmental Assessment Agency and Utrecht University with my close involvement, is outlined.

Climate change constitutes the main example of a public policy issue in which the use of computer simulation is hotly contested. Section II delves more deeply into this case. Chapter 5 describes the practice of climate simulation, while Chapter 6 discusses the uncertainties in climate simulation. Chapter 7 deals with the assessment of climate-simulation uncertainty for policy advice, with particular attention paid to the IPCC, a scientific assessment body of the United Nations that provides advice to policymakers on climate-change issues.

Chapter 8 presents the conclusions of this study by answering the research questions posed in this introduction.

Section I

Simulation Practice, Uncertainty, and Policy Advice

2

The Practice of Scientific Simulation

2.1 Introduction

Simulationists, in their daily practice, draw on a vast array of heterogeneous resources, such as mathematical models and computers (hardware and software); input data needed for running the models; results of experiments or observations (for preparing the input data and for comparison with the output); general theories (for basing models on and for comparison with the output); skills and methodologies for developing and evaluating simulations; social relations within all kinds of institutions, be it simulation laboratories,[*] universities, government and business research institutes, scientific disciplines, professional societies, peer review systems, and the like or society at large. The list of elements of scientific simulation practice can be drawn up, extended, and refined in an unlimited number of ways.[†]

For analysing epistemological and methodological aspects of simulation, I argue that the activities of simulationists can be conceptually subdivided into four main types: (1) formulating the mathematical model, (2) preparing the model inputs, (3) implementing and running the model, and (4) processing the data and interpreting them.[‡] These four types of activities refer to four epistemologically distinct elements in simulation practice: (1) the conceptual and mathematical model; (2) model inputs; (3) the technical model implementation; and (4) processed output data and their interpretation. Clearly, not all resources and activities in scientific simulation practice are captured by these categorisations. In as far as other resources and activities impinge on the four main types of activities, they are taken into account in our description of simulation practice.

---

[*] The term *simulation laboratory* is introduced in the following discussion.

[†] Hence lists enumerating the elements of scientific practice typically end with 'etc.', as is discussed by Hacking (1992: 31–32) and Hon (2003: 181–182).

[‡] The present approach to a conceptualisation of scientific practice derives from Hon (1989), who applied it to the practice of experimentation and termed the different types of activity 'stages'. Although his term *stages* was not intended to imply that the activities belonging to the different stages are actually carried out in a consecutive order (typically many iterations take place; see also page 21, second note), I do not adopt his term to avoid confusion.

In the present chapter, first, the practice of scientific simulation is described as a 'laboratory,' and the four main types of activity in this practice are introduced (Section 2.2). Second, the corresponding elements are studied more closely, and several philosophical issues are addressed, such as the relation between general theory and models; the proximity of simulation and material experiments; the reproducibility of simulation; and the role of visualisation in scientific understanding (Section 2.3). Subsequently, the plurality of methodologies for developing and evaluating simulations (Section 2.4) and the plurality of values (Section 2.5) in scientific simulation practice are examined. Finally, the similarities and differences between simulation and experimentation practice are addressed (Section 2.6).

2.2 The Simulation Laboratory

For describing the structure of simulation practice, it is helpful to consider simulation science to be a 'laboratory' science and use some of the relevant insights from philosophical and social studies of laboratory work. Following Ian Hacking (1992: 33–34), scientific 'laboratories' are here defined as sites, typically within universities and government or business institutions, where scientific knowledge is produced and that satisfy the following two conditions:

(L1) the claims made by the scientists working in the laboratory refer primarily to phenomena 'created' there (as opposed to observational practice);

(L2) the laboratory practice has become stabilised and institutionalised (as opposed to scientific practices that have not yet 'come of age').

The notion of 'laboratory' includes the objects, the ideas, the procedures, the people, the buildings, the institutions, and so on involved in doing scientific work. I propose to generalise this notion of the 'laboratory' from experimental practice (for which Hacking reserved the term) to simulation practice. I argue that phenomena are 'created', albeit digitally, within the simulation laboratory (satisfying L1), and that the levels of stabilisation and institutionalisation of simulation practices are sufficiently high for satisfying L2 in many instances of scientific simulation.*

---

* This does not imply that simulation uncertainties, the main topic of this study, are small or unimportant. It just means that several of the uncertainties that simulationists face—both methodological and institutional—are reduced, in the sense that there is a community of scientists who share the same methodology. In addition, there are some general statistical methods and procedures for dealing with uncertainty that are widely shared among simulationists. However, despite the stability of simulation practice, I argue in Chapter 3 that the application of statistical methods is not sufficient for adequately dealing with uncertainty.

With respect to Hacking's condition L1, there is obviously a fundamental difference between experimental and simulation practice: In simulation practice, we do not *physically* bring nature into the laboratory, and thus we do not have to control nature through material intervention and create a 'closed' system (see Radder 1988: 63–69) that is insensitive to influences external to the phenomenon of interest. Instead, in scientific simulation we bring a mathematical representation of nature into the laboratory, and control of external influences is not really an important issue; such control—that is, creating a closed system—is obtained by stipulation. Still, there is an important similarity between experimental and simulation practice. Both practices involve extrapolations to the outside world. Simulationists aim to bring the outside world into the laboratory first by way of mathematical representation and subsequently extrapolate the results to the world outside.* In the experimental laboratory, extrapolation to the outside world takes place as well when experimentalists use apparatus as models of systems in the world. The difference between the experimental laboratory and the simulation laboratory is that in experiments the processes studied inside the laboratory are supposed to be of the same material kind as the processes occurring outside (see, e.g., Harré 2003: 26–32, who distinguishes between two classes of experimental models: 'domesticated models' of natural systems and 'Bohrian apparatus-world complexes').† It is this similarity of having a model, either material or theoretical, of the outside world, as a primary target of study and therewith 'creating' phenomena inside the scientific site that leads to the proposition that simulation practice satisfies condition L1 for being a laboratory practice.

Hacking's condition L2 refers to the stabilisation and institutionalisation of a laboratory practice. The stabilisation of simulation-laboratory practice primarily depends on whether coherence can be created and maintained among the vast array of heterogeneous resources (mathematical models, computers, input data, results of experiments and observations, general theories, skills, methodologies, social relations, etc.). An important sign of both stabilisation and institutionalisation is that many simulations are done remotely, that is, on a computer that is not on the simulationist's desktop. In those cases, simulationists often do not even know what the remote computer (e.g., a supercomputer) looks like, let alone how the hardware works; simulationists can use the remote computer if they can adequately act within the

---

* The material processes occurring in the simulation laboratory are of course very different from the material processes that are represented: The electrical processes taking place inside the computer in no way materially resemble the phenomena outside the simulation laboratory.

† The Bohrian apparatus-world complexes have the characteristic that the phenomena created in the laboratory do not occur in nature. Back inference from these phenomena to nature outside the laboratory is problematic since the apparatus contribute to the form and qualities of the phenomena. Still, the dispositions of nature actualised in Bohrian apparatus 'permit limited inferences from what is displayed in Bohrian artefacts to the causal powers of nature' (Harré 2003: 38).

software environment of the remote computer. The software must be highly standardised for many simulationists to be able to use it. If we consider such a remote computer as part of the simulation laboratory, a picture arises of globally connected networks of simulation laboratories, collaborating in an institutionalised setting. For example, different simulation laboratories may use the same computer: One supercomputer may be used simultaneously for running simulations related to atmospheric flows, molecular dynamics, particle physics, pharmaceutics, ecology, and the like without the simulation practices showing more overlap than a sharing of the same hardware and system software (including programming languages) and several simulation techniques that have become highly credible as 'reliable techniques or reasonable assumptions' across fields (Winsberg 2003: 122). The simulation practices differ in mathematical models, input data, results of experiments or observations, general theories, social relations, etc. Still, even though their practices are different, the present study argues that an important set of skills and methodologies to develop and evaluate simulations plays some role in all scientific simulation practices.[*]

That simulation practices have become institutionalised, separately—but usually not cut off—from theoretical, experimental, or observational practices can also be concluded from the presence of highly rated journals with 'computational' in their titles[†] and the fact that many research groups use the phrase 'simulation laboratory' in their names. Most simulationists, however, consider themselves in the first place physicists, chemists, biologists, atmospheric scientists, geologists, earth system scientists, and so on. And they see their simulation practices, with identifiable cultures that are distinct from the cultures of theoretical, experimental, and observational practices, as inextricably linked with these other practices. Thus, one can say that there are divisions of labour within science between different types of scientific practices that are bound together by subject matter (Galison 1996; Galison 1997: Ch. 8).

Divisions of labour can also be observed within simulation practices, especially in 'big science' practices (e.g., climate modelling). Within large simulation research groups, there are people who have specialised in hardware and system software; others who write scientific programs according to given specifications; and again others who are involved in defining and distributing 'frozen' versions of the programs. Furthermore, we have scientists who specify what the programs should do and who uses the programs to learn from them. Finally, there are numerical mathematicians who specialise in

---

[*] More evidence for the stabilisation of simulation practice is provided in Section 2.3.3.

[†] The number of journals in the Thomson Scientific Master Journal List (http://science.thomsonreuters.com/mjl) containing the word *computational* in their titles was 37 on 7 July 2006 (the numbers for *simulation* and *computation* were 10 and 16, respectively). Five years later, on 5 June 2011, these numbers had significantly increased: 55 for *computational*, 14 for *simulation*, and 19 for *computation*.

crafting numerical algorithms that may be used as standardised software packages linked to the programs. This division of labour will not be found in smaller laboratories; there the roles usually overlap in individuals. Thus, at one extreme there exist scientists who themselves build special-purpose computers, write their own system software, master all the necessary numerical mathematics, write the scientific programs, learn from these programs, and publish articles on the scientific subject studied. Such scientists are rare, however. At the other extreme, all these tasks are performed by different groups of persons. This is also a rare situation. The most common situation is that some of the roles overlap in individuals, with individuals having different levels of expertise with respect to these roles. Some simulationists know a great deal about computer technology and system software, and others are more focused on the scientific context (they may even not want to call themselves 'simulationists'). Within the latter group, variation in the scientists' skills in numerical mathematics and scientific programming exists.

It is thus possible to generalise Hacking's notion of the laboratory to scientific simulation in the sense that conditions L1 and L2 are fulfilled in simulation practice. We can now take the analogy with experimental laboratory practice one step further and investigate whether there are analogons in simulation-laboratory practice of the elements that Hacking distinguishes in experimental laboratory practices. Hacking (1992) identifies three main categories of elements of laboratory practice: 'ideas', 'things', and 'marks'. I propose that in simulation, the main ideas simulationists work with are conceptual and mathematical models; the main things are the implementation of these models on computers (the 'technical model implementations'); and the main marks are input data (often derived from other practices, from experiments, observations, or simulation) for and output data from the technical model.* To these four elements correspond four main types of activities: (1) formulating the mathematical model, (2) preparing the model inputs, (3) implementing and running the model, and (4) processing the data and interpreting them. Evaluative and reconstructive activities concerning the mathematical model, the model inputs, the model implementation, and processing and interpretation of the data are all taken to belong to one of the four main types of activities; they may lead to revisions.† The activities in simulation, the corresponding elements, and Hacking's categories are shown in Table 2.1. In the next section, I discuss consecutively each of the four elements of scientific simulation practice.

---

* Input and output marks are here considered as separate elements.
† As Nickels argues (1988: 33), '[s]cience transforms itself by more or less continuously reworking its previous results and techniques'. A 'multipass' account of scientific practice more adequately captures its dynamic than a 'one-pass' account.

TABLE 2.1

The Four Activities and Elements of Simulation and Their Correspondence to Hacking's Categories

Activities	Elements	Categories
Formulating the mathematical model	Conceptual and mathematical model (structure and parameters)	Ideas
Preparing the model inputs	Model inputs (input data, input scenarios)	Marks (input)
Implementing and running the model	Technical model implementation (software and hardware)	Things
Processing the data and interpreting them	Processed output data and their interpretation	Marks (output)

2.3 Elements of Simulation-Laboratory Practice

2.3.1 Conceptual and Mathematical Model: General Theory, Models, and Parameterisations

The central idea element of a simulation is the mathematical model. The conceptual model associated with the mathematical model offers its interpretation and determines what is considered inside and outside the boundaries of the system under study. Other idea elements distinguished by Hacking that play a role in simulation practice are the questions that drive the development, evaluation, and application of simulation models; the hypotheses that simulationists are testing; and general theory. But, how do these different idea elements relate to each other? Here I focus on the relation between general theories and models.

Models can be based on general theories, but models are seldom derived from such theories alone. Besides general theory, the model construction process involves the combination of a heterogeneous set of elements such as fundamental principles, idealisations, approximations, mathematical concepts and techniques, metaphors, analogies, stylised facts, and empirical data (Humphreys 1995: 502; Boumans 1999: 94). Individual simulationists have to make many choices in model construction and often have considerable leeway in making these choices. In the mathematical formulation of the model, a distinction can be made between the 'model structure' (the mathematical form of the equations) and the 'model parameters' (the constants in the mathematical equations). According to Paul Humphreys (1995: 509), theories can typically legitimate model structure but not model parameters. The latter then need to be determined from empirical data. This often happens by trial and error and comparison of model outputs with real-world data.[*]

---

[*] Some models allow for the specification of the model parameters as model input, which facilitates the variation of these parameters.

Using Keller's typology of simulation (see Chapter 1), we can distinguish three ways in which the conceptual and mathematical models can relate to general theory: (1) the model only contains *mathematical* approximations to general theory; (2) the model also contains *conceptual* approximations to general theory; and (3) the model does not relate to any general theory. An example of simulation that belongs in the first category is the technique of direct numerical simulation of turbulence that is widely used in fluid dynamics. Fluid dynamics is an archetypical domain of science in which analytical mathematical methods fall short. The fundamental mathematical laws (general theory) that are used to describe the dynamics of fluid flows, the Navier–Stokes equations, are intractable for turbulent flows. Given enough computational resources and a not-too-high intensity of the turbulent flow, all relevant details of turbulent laboratory flows can now be modelled by direct numerical simulation. Only approximations of a mathematical nature are made in this kind of simulation. The second type of simulation is the most common type in the physical sciences. In fluid dynamics, for instance, the Navier–Stokes equations are also applied in other types of fluid-dynamical simulation models (i.e., atmospheric models). In those models, conceptual approximations to the Navier–Stokes equations also need to be made due to the high turbulence intensity. As was argued by Adam Morton, in order to retain satisfactory theoretical quality, simulationists are typically required

> to give a heuristic physical argument that the simplification or idealization should not affect some category of consequences for the states of the system too drastically. (Morton 1993: 661–662)

To give an example, a typical approximation is the neglect of processes occurring at scales smaller than a certain limit; this is argued for on physical grounds.* An example of the third category is the use of cellular automata in biological research described by Keller (2003).

Let us here elaborate briefly on the epistemological difference between general theories and models. Weinert (1999: 307) argues that 'the constraint structure imposed on models is not the same as the constraint structure which applies to theories'. The distinction roughly boils down to the types of constraints that have to be satisfied by general theories as compared with models. General theories have justificatory functions. For instance, they are means for deriving empirical laws. For models, empirical adequacy is more important. Models do not necessarily depend on the existence of an underlying general theory. Furthermore, the principles used in models may be

---

* Some philosophers of science have suggested that many approximations in models are purely mathematical, leaving the basic physical assumptions intact (see, e.g., Redhead 1980; Koperski 1998). One of the examples given by Koperski (1998: 629) is the approximated form of the Navier–Stokes equations that is used for simulation in fluid dynamics. It is argued here, however, that mathematical and physical approximations go hand in hand in this case: The mathematical approximations are justified by physical arguments.

ad hoc (that is, having no independent theoretical or empirical basis); and models may be embedded in false general theories while retaining a relative independence of the fate of these theories (Weinert 1999: 320–321). Models may evolve and acquire the status of general theory, or they may give rise to and be replaced by general theory.*

However, the boundary between models and general theories is not as sharp as Weinert suggests. The distinction must be considered to be a relative one. This can once again be illustrated by means of the Navier–Stokes equations of fluid dynamics. These equations, considered as a general theory in fluid dynamics, can themselves also be considered as a model from the viewpoint of one of our most fundamental physical theories, statistical mechanics (see Cartwright 1983: 63). The Navier–Stokes equations can only describe fluid flow approximately and fail at very small scales or under extreme conditions. Real fluids made out of mixtures of discrete molecules and other material, such as suspended particles and dissolved gases, will produce different results from the continuous and homogeneous fluids modelled by the Navier–Stokes equations. In some of those conditions, statistical mechanics may be a more appropriate approach. Still, the Navier–Stokes equations have a wide range of applicability and can be derived from theoretical constraints such as conservation of mass, momentum, and energy, while making only a small number of assumptions. In geophysical fluid dynamics the Navier–Stokes equations are themselves considered a general theory that has to be approximated by models in order to perform calculations. A theory can thus be considered to be a general theory in one context and to be a model in another, from the perspective of more fundamental theories and depending on the domain of application.† Although in each of those contexts a distinction between general theories and models can be maintained, the interpretation of the separate constraints for theories and models identified by Weinert will differ from one context to another.

---

* A prime example of a model, often discussed by philosophers of science, is Bohr's 1913 model of the atom. Bohr's inconsistent model led to the consistent theory of quantum mechanics (after the model had been transformed). Note that Morrison and Morgan (1999: 36) express their concern that from the way models have been treated in the philosophy of science people could erroneously conclude that all models must be regarded as 'preliminary theories'.

† Cartwright's (1983) distinction between fundamental and phenomenological laws should thus also be relativised. Cartwright (1999: 242) herself later revised her view on the nature of the distinction: 'I shall call these [phenomenological laws] *representative models*. This is a departure from the terminology I have used before. In *How the Laws of Physics Lie* (1983) I called these models *phenomenological* to stress the distance between fundamental theory and theory-motivated models that are accurate to the phenomena. But *How the Laws of Physics Lie* supposed, as does the semantic view [which considers theories as sets of models], that the theory itself in its abstract formulation supplies us with models to represent the world. They just do not represent it all that accurately. ... Here I want to argue for a different kind of separation: These theories in physics do not generally represent what happens in the world; only models represent in this way, and the models that do so are not already part of any theory.' But she stills retains the distinction between fundamental theory and representative models; hence, my criticism that this distinction is a relative one still applies.

I now turn to the issue of 'parameterisation' in modelling as an illustration of the fact that the mathematical models used in simulation are most often not fully derived from general theory. The examples here come from atmospheric modelling. Due to the sheer size and complexity of the atmosphere, the direct application of general theory such as the Navier–Stokes equations of fluid dynamics to practical atmospheric problems is unfeasible, even if we use the fastest computers that exist. Forecast models* of the atmosphere therefore contain a number of 'parameterisations' of processes that cannot be directly simulated. A *parameterisation* is a mathematical model that calculates the net effects of these 'unresolved' processes on the processes that *are* directly calculated in the forecast model (the 'resolved' processes). There are several ways to arrive at parameterisations. From case study measurements of the unresolved processes, one can determine the statistics that describe the net effect of these processes on the variables in the forecast model. The statistical relationships obtained can be included in the forecast model. Alternatively, deterministic models (also considered to be parameterisations then) can be developed that simulate the statistics directly. These models can again be included in the forecast model.

I give an illustration of how I derived a parameterisation—in the form of a model that directly calculates statistics—for the influence of convective atmospheric boundary layer[†] turbulence on the chemical reaction rates calculated in regional or global chemistry–transport models, that is, large-scale air pollution models. The horizontal grid size of these large-scale models is typically a few hundred kilometres, while the scale of the turbulence processes is a few hundreds metres at most. Thus the turbulence processes cannot be resolved in the large-scale air pollution models. My solution was to propose an 'updraught–downdraught model' or 'mass-flux model' (Petersen et al. 1999) as a parameterisation that could be included—in the simplified form of Petersen and Holtslag (1999)—in the large-scale models. The conceptual model is shown in the diagram of Figure 2.1. All the air that moves up through turbulent processes within a grid cell of a large-scale model is conceptually lumped together in one 'updraught' and vice versa the air that moves down is lumped together in one 'downdraught'. I will not enter into the details of the processes and of the mathematical model. The only thing that I want to point out here is that the model contains a parameter κ that connects the effect of turbulence at smaller scales than the updraught and downdraught to the effect that is explicitly modelled by the updraught–downdraught separation.

---

* In this context, *forecast models* are models that 'predict the deterministic evolution of the atmosphere or some macroscopic portion of it', for example, numerical weather prediction models (Randall and Wielicki 1997: 400).

† The atmospheric boundary layer, also called the planetary boundary layer, is the turbulent layer at the bottom of atmosphere (near the planetary boundary—the earth's surface), which varies in height between approximately 100 and 1,500 m. If the boundary layer is convective, plumes rise and sink.

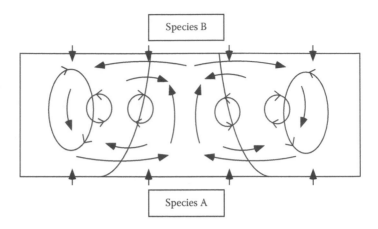

FIGURE 2.1

Two conceptions of turbulent mixing in the convective atmospheric boundary layer shown together for a given volume of about 1 km in the vertical dimension and several kilometres in the horizontal dimension: (1) schematic view based on the updraught–downdraught decomposition (the two plumes are separated by solid lines, with all air that rises conceptually lumped together in one plume—shown in the middle here—and all other air sinking in the other plume, as shown to the left and right of the middle plume here; the full arrows in the plumes indicate the average flow); and (2) the presence of eddies at all scales (designated by some larger and smaller eddies with open arrows). Also shown are the boundary conditions for a simple chemistry case. Chemical species A is introduced at the surface, and chemical species B is injected at the top of the boundary layer. (From Petersen, A.C. (1999b), Convection and chemistry in the atmospheric boundary layer, PhD dissertation, Utrecht, the Netherlands: Utrecht University, 10.)

Here we arrive at the source of the controversy about the soundness of simulation mentioned in Chapter 1. To determine κ, instead of using observations (which are nearly impossible to perform), I used a three-dimensional (3-D) large-eddy simulation (LES) model that simulates the turbulence processes in the convective atmospheric boundary layer.[*] From simulations, I concluded that $\kappa = 0.25$, and I claimed that assuming this value would result in an estimate of the net effect of subgrid chemical reactions that was accurate within a factor of two when averaged over the whole boundary layer in a grid cell of a large-scale model.

The practice of parameterisation is widespread in scientific simulation and typically involves model calibration (e.g., Janssen and Heuberger 1995). For the types of simulation that are of central interest to this study, deterministic models in the geophysical sciences, the ideal of many practicing simulationists is realism, that is, providing a realistic representation with their parameterisations. This type of simulation corresponds with Keller's second type of simulation. In my case, the updraught-and-downdraught decomposition refers to processes that can be measured if enough effort is expended.

---

[*] Thus, a parameterisation within a simulation can itself consist of, or be derived from, another simulation.

However, I used LES as a substitute for measurements to be able to study idealised cases under widely varying but controlled conditions that allowed me to determine the scope of the parameterisation efficiently. The development of simple models by comparing them with complex models that give a detailed description of reality has been observed by Hans von Storch (2001) to be a ubiquitous feature of environmental science.[*]

Parameterisations also occur in simulations belonging to Keller's third type of simulation. In this type of simulation, phenoma are simulated 'for which no general theory exists and for which only rudimentary indications of the underlying dynamics of interaction are available' (Keller 2003: 202). Fritz Rohrlich (1991: 512–514) gives the example of a model called 'stochastic self-propagating star formation' that simulates the evolution of a spiral galaxy.[†] This model contains various parameters that can be determined from observation, such as the rotational speed and the radius of the galaxy. But it also has adjustable parameters such as the probability of star formation within a given time period. As Rohrlich (1991: 514) puts it, the question is 'whether trial and error will yield a final appearance of the simulated galaxy that is close enough to what is observed'. Visualisation techniques make it possible to compare the model results with real data visually (see Section 2.3.4).

Thus, a varied picture arises of the role of general theory in simulation practice. At the one extreme, simulation models rely heavily on general theory. At the other extreme, simulation models involve no general theory. Many simulation models lie between these extremes. Parameterisations (which are mathematical models within simulation models that represent the net effect of unresolved processes on resolved processes) often contain adjustable parameters whose setting is not based on theory.

2.3.2 Model Inputs: Proximity of Simulation and Material Experimentation

Simulation models require input data that specify initial and boundary conditions needed for the numerical integration of mathematical models. These input data may come from other simulation models, from experiments and observations, or from 'educated guesses' or they may be idealisations.[‡]

For many simulation models, experimental or observational work is involved in the preparation of the input data. Numerical weather prediction

---

[*] von Storch (2001) gives a positive judgement of this situation. Both simple and complex models are needed. He calls the simple models 'cognitive models' and the complex models 'quasi-realistic models'. The latter play the role of surrogate reality, and they are themselves not considered useful by von Storch for generating understanding.
[†] This is an example of the application of cellular-automata models outside biology, the discipline to which Keller (2003) restricts her discussion of the third type of simulation.
[‡] An example of the latter are the idealised input data that I used for the simple and complex models in my turbulence–chemistry simulations; in many simulations, the fluxes of chemical species into the boundary layer were kept constant.

models, for example, require an extensive amount of input data from weather stations, balloons, and satellites to be able to accurately predict the weather a few days ahead. The accuracy of the predictions is dependent on the accuracy and completeness of the input data.* Another example is the production of historical global data sets of atmospheric flows, temperature, cloud cover, and so on that are used to compare with climate model output. These data sets are the result of an incorporation of surface, weather balloon, and satellite data into a four-dimensional (4-D) data assimilation system† that uses a numerical weather prediction model to produce global uniformly gridded data (Edwards 1999: 451). Simulation practices can thus be intrinsically connected with observational practices. By feeding real-world data into simulation models, these models can be used for measurement—and they can even be considered as part of the measurement apparatus (see, e.g., Norton and Suppe 2001; Morgan 2003). For instance, models are involved in satellite measurements of particular quantities in the atmosphere, such as the vertical temperature or ozone concentration profile (Norton and Suppe 2001). The satellites can only measure radiation; atmospheric radiative transfer models are needed to infer temperature and ozone concentrations.

Similar connections can be found in experimental practice. For example, the U.S. Oak Ridge National Laboratory has proposed starting a Virtual Human Project, which can use digitised anatomical images from the U.S. National Library of Medicine's Visible Human Project and can perform model experiments on humans (e.g., testing the physiological and biochemical effects of particular drugs).‡ Such a project would have a scale larger than the Human Genome Project. A second, smaller-scale example is from biomechanics: Three-dimensional simulations have been developed that can be used to measure the strength of specific bones based on input produced by experimental work in the laboratory (the bones are sliced, and the slices are subsequently scanned into the computer). A model based on a 'conventionally accepted (tried and tested in the applied domain) mathematical version of the laws of mechanics' (Morgan 2003: 222) is used to simulate the breaking of the bones. Morgan (2003: 224), who analysed this example of computer-aided measurement, calls it 'virtually an experiment' because of the proximity of the input data to the real material world. She considers the input data to be 'semimaterial' (Morgan 2003: 221–224) and distinguishes the simulation of

---

* Clearly, the accuracy of simulation results also depends on the reliability of the model (see Chapter 3).
† Four-dimensional (4-D) data assimilation is a statistical technique for bringing time-dependent 3-D model variables as close as possible to observations taken at different places and different times.
‡ For information on the Oak Ridge National Laboratory Visible Human Project, see http://www.ornl.gov/sci/virtualhuman. The National Library of Medicine's Visible Human Project can be accessed at http://www.nlm.nih.gov/research/visible/visible_gallery.html.

the breaking of a semimaterial bone from simulations in which a computer-generated stylised bone is used (which she calls 'virtual experiments').*

Morgan argues that the simulated breaking of a semimaterial bone is 'more like an experiment on a material object', and the simulated breaking of a stylised bone is 'more like an experiment on a mathematical model' (Morgan 2003: 224). This formulation may give rise to a misunderstanding, presumably not intended by Morgan. Obviously, what Morgan calls 'semimaterial objects' also need to be mathematically represented as input data to the model, just as is the case for the stylised bone. The difference that Morgan wants to stress is that the verisimilitude of one set of input data is higher than that of another set. Thus the only sense in which we can call virtually experiments closer to material experiments is that the mathematical input of the simulation resembles more closely the material input of the material experiment. It depends on the reliability of the model used to simulate the intervention whether the results of simulations remain close to those of material experiments.

The preparation of input data for simulation constitutes one important locus of interaction between the different practices of simulation and experimentation. To what extent the use of such input data brings simulation closer to a material experiment depends also on the mathematical model that is used. Section 3.6.2 elaborates this question. Still, we can conclude that the accuracy of a simulation depends partly on the accuracy of the input data.

2.3.3 Technical Model Implementation: The Reproducibility of Simulation

The main *thing* element in simulation-laboratory practice is the 'technical model implementation', which consists of all hardware and software involved in implementing and running the mathematical model. The thing elements distinguished by Hacking—targets, sources of modification (of the targets), detectors, tools, and data generators—are all identifiable within the technical model implementation. Simulationists craft these elements as their tools for studying the behaviour of the mathematical model, and since they assume a representational relationship to the systems of interest in the outside world, the technical model implementation is used as a tool to study those systems.

The *targets* in the simulation laboratory are the digital representations of the systems of interest as computer programs. These computer programs represent both the state of the simulated systems (through 'main program variables' stored in the computer's memory) and the dynamical behaviour

---

* The stylised bone has 'a structure that begins as a simple 3-D grid of internal squares. The individual side elements within the grid are given assorted widths based on averages of measurements of internal strut widths (taken from a number of real cow bones) and are gently angled in relation to each other by use of a random-assignment process' (Morgan 2003: 222).

of the systems (through 'endogenous' changes—that is, changes related to the internal dynamics of the system—in the main program variables calculated by program routines). For numerical weather prediction models, for example, the main program variables represent the state of the atmosphere (e.g., temperature, humidity, winds, etc.), and some of the program routines represent the fluid-dynamical changes in these variables. Sources of modification, which are also implemented as program routines, allow for a direct external influence on the main program variables. With respect to air pollution models, for example, one can think of the emission of pollutant gases, which is implemented in the program as increases in pollutant-gas concentrations. Furthermore, the state of the system can be 'measured' by *detectors*, which are program routines that have access to the main program variables—not to change these variables but to store them or perform calculations on them (such as calculating averages). Several other tools can also be identified in simulation-laboratory practice. These range from software components other than the simulation model implementation (such as text editors, program compilers, or program debuggers) to technological aspects of computing (such as keyboards, mice, computer screens, or printers). Finally, *data generators*' are routines that produce output (data) which can be used for further processing (such as making plots of main program variables).

The skills needed for dealing with the thing elements in simulation have changed over time. In the early days, more manual labour was involved in writing programs and operating the computer than is necessary currently for solving comparable mathematical problems. The differences are related to changes in hardware and in software. For instance, the first computers continually broke down, and scientists cooperated with technicians in performing repairs. Furthermore, the ease of use of today's user interfaces was lacking: The earliest computers did not have a keyboard, screen, or mouse. As a consequence, the interactivity of scientific computing—the ease and speed with which one can make changes in programs or input data—was quite low in the early days. The changes in software are mainly related to the development of programming languages. Originally, programmers had to write their programs directly in 'machine code', a list of instructions that could be directly executed by the hardware (e.g., 'store the value X in register Y', or 'add register Y and register Z and put the outcome in register Y'). In the course of time, higher-level programming languages evolved (such as FORTRAN or C), which allowed the programmer to specify the operations in much less detail, at a higher level of abstraction.* It became possible, for instance, to define 'procedures' (or 'subroutines') within computer programs, a part of the code that could be reused efficiently. There is still a continuing evolution of programming languages, by which even higher levels of abstraction are reached, typically facilitating Keller's third type of simula-

---

* The first FORTRAN compiler was delivered in 1957 and the first C compiler in 1973.

tion.* Present-day programming languages allow programs to be written that are too difficult to be written directly in machine code—not because this is theoretically impossible (ultimately, all software is run as machine code) but because the capacities of humans to oversee and comprehend programs at the machine code level are limited (that is, the level of detail is too high). As far as the handling of the computer equipment is concerned, simulationists have to know how to handle input and output files and system software. They must be able to handle 'compilers' (software that translates high-level programming language into machine code executable by the computer) and debuggers (software that helps the programmer in tracing 'mistakes' made in writing or typing the computer program; such mistakes in the computer code are called 'bugs'†). There are also special skills involved in dealing with the mathematical level of the simulation model. The equations of the model, which can often be written on paper in an analytic form, have to be implemented in a numerical form. This involves discretisation and approximation.

In 'big science' simulations (i.e., in ecological or climate science), the computer simulation programs can become very large, containing thousands of programming lines. Many of the large scientific computer programs in current use have been built by groups of scientists. Scientists who are new to these programs have to put much effort in understanding what is happening where in the program and why the program was written in a particular way. This is why it is a sign of good modelling practice if a comprehensive explanatory commentary is included in the text of the computer program (this inclusion is facilitated by high-level programming languages). If such large programs also use large amounts of input data, it can become a tedious task to keep track of the input to the model. This is also a case for which documentation is essential. However, in practice, not all scientific simulations are well documented. Master–apprentice relationships are indispensable for transferring the skills needed to run or further develop many scientific simulations. Furthermore, scientists keep making changes in the programs and their input, so they also have to keep track of these changes, as well as of the impacts of these changes on the model output.

Part of the stability of scientific simulation practice depends on the reproducibility of simulation at the level of the technical model implementation.‡ Peter Galison (1996: 140–141) concludes from the history of scientific simulation that its reproducibility was problematic in the early days of scientific simulation. Simulations were hardly 'transportable' to other simulationists, which was considered a 'bottleneck' to 'delocalize simulations and computer-analysis programs' (Galison 1996: 140). Galison records that three

---

* In 'object-oriented' programming languages, for instance, the concepts of 'classes' and 'instances' of objects allow for a new kind of modelling approach, facilitating easy implementation (e.g., of cellular automata).
† Bugs are errors at the stage of model implementation, not at the stage of model formulation.
‡ The 'norm of reproducibility' has been shown to play a significant role in the stability of experimental practice (see Radder 1996: 9–44, 119–135).

solutions to this situation were proposed: (1) to publish computer programs openly, (2) to make programs 'portable' from simulationist to simulationist by using universal programming languages and physically distributing data tapes, and (3) to consider 'modular' programming and the use of standardised subroutines, an important goal for scientific programmers. All three solutions were partially realised in the second half of the 20th century (Galison 1996: 141), giving current simulation practice a definite 'nonlocality', although not universality.

The reproducibility of simulation is greatly facilitated by standardisation in both hardware and software. The level of standardisation in computer hardware is now high. Computer technologies can usually be easily replaced when broken or can be set up elsewhere—provided that locally a minimal level of skills and the necessary hardware and system software are available (e.g., spare parts or access to a centralised file system through a network). Even though some computing technology (i.e., the technology used in supercomputers) may require carefully maintained ambient conditions, this is typically not the business of simulationists. They use the computer as it is and are not much interested in the construction of new, experimental computers.[*] Also a part of the software used in simulation is highly standardised. Work has been done to increase the level of standardisation (e.g., the syntaxes of programming languages have been defined through international standard-setting agreements). We currently have a small number of standard operating systems (e.g., Windows, Mac OS, Linux, UNIX); programming languages (e.g., FORTRAN, C); and general simulation-cum-visualisation software packages (e.g., MATLAB, IDL).

However, the computer programs themselves that are written by the simulationists differ widely in their level of standardisation (e.g., modular programming style, use of standard numerical mathematical routines).[†] Programming language compilers may be used that offer additional features that deviate from the standard. 'Good' simulationists are aware of the advantages of sticking to the standard and will not make use of such features; this means that it may be expected—although it cannot be guaranteed—that a certain program that works on one type of computer will also run on another type, provided that a similar compiler is also available for the other type of computer. However, in practice such portability is not realised for most simulation programs since many simulationists do not stick to the standard. Furthermore, virtually all simulation programs contain local elements, often understandable (logically, mathematically, and conceptually) only to the simulationists who wrote the programs and to their apprentices. All this poses limits to the reproducibility of simulation and introduces requirements for simulationists

---

[*] Notably, some simulationists *are* building new computers, but this currently is the exception, not the rule.

[†] Although some simulationists are involved in the development of new software techniques and numerical calculation methods, that is, are involved in the continuing process of standardisation, they are a minority.

to develop specific skills to work within a particular scientific simulation practice, in addition to being able to work with computers in general.

Let us look at the issue of the reproducibility of simulation in somewhat more detail. Hans Radder distinguishes three types of reproducibility of experiment. Here, I generalise his notion of reproducibility to the reproducibility of simulation. The types can be specified based on the different roles played by, on the one hand, the technical model implementation (part of the 'material realisation', in Radder's terms, which relates to the manipulation of things) and, on the other hand, the conceptual and mathematical model and the results from the simulation (together called 'theoretical interpretation' by Radder).* The three types of reproducibility of simulation that I distinguish are (1) reproducibility of model runs with the same technical model implementation; (2) reproducibility of the simulation under the same theoretical interpretation (conceptual and mathematical model and result), possibly with another technical model implementation; and (3) reproducibility of the result with a different simulation (different conceptual, mathematical, and technical models) or with an experiment, also called 'replicability'. In this section, the first two types of reproducibility are discussed. The norm of replicability is addressed in Sections 2.4 and 3.6.3.

The material realisation of a simulation refers to everything that happens at the level of things, that is, the model implementation on the computer, and the way simulationists handle these things.† The processes within digital computers are supposed to proceed in an orderly rule-based manner, and the 'natural variability' within them can therefore be said to be negligible by design. Thus when programmers work with computers, they normally do not have to deal with 'noise' in a computer's memory or its processors. Of course, designers sometimes make errors: Hardware and software designs may be flawed, and things can therefore go wrong during the runtime of a simulation. A famous example of a hardware design error is provided by Intel's Pentium chip (Markoff 1994). In June 1994, Intel received information from scientists using the Pentium chip for scientific calculations that for certain calculations the chip was exact to only 5 digits, not 16, as it was supposed—and specified—to be. Intel found out that the error occurred because of an omission in the translation of a formula into computer hardware. It was corrected later that year by adding several dozen transistors to the chip. Aside from the possibility of being flawed in design, hardware may also

---

* The theoretical interpretation $p \rightarrow q$ consists of the theoretical result (q)—a proposition—and other theoretical descriptions (p) that all play a role in inferring the result.
† The skills needed to carry out simulation studies have been discussed. They mainly reside at the theoretical level and less at the material level of interacting with the computer. Thus there is a hypothetical but realisable possibility that all physical interactions with the computer (typing on the keyboard, moving the mouse, etc.) are performed by laypersons receiving instructions in ordinary language from the simulationist (cf. Radder 1988, who introduced the concept of laypersons receiving instructions in ordinary language in his analysis of the material realisation of experiments).

occasionally break down. Although there is no absolute guarantee that a program will always run without failure, one may typically assume that repeating a run of the same program on the same computer with the same software environment gives exactly the same results.* Once a model 'works', it will keep working. Still, we should not conclude that the norm of the reproducibility of the material realisation of a simulation (being able to perform the same runs with the same model) is always upheld in practice. The reason is that the 'same computer' and the 'same software environment' may not be available at a later date.

The main function of the second type of reproducibility, reproducibility of the simulation under the same theoretical interpretation (conceptual and mathematical model and result) with a possibly different technical model implementation, is to be able to distribute the models to different computers.† In this case, changes often have to be made within the simulation program so that it can be run in different software environments (using different compilers, for instance). This notion of reproducibility demands that after making such small changes to the existing technical model implementation, or alternatively (which sometimes happens), after entirely rebuilding the technical model implementation, the results are compared. This provides a check on the effects of numerical approximations and mistakes in technical model implementations.

Examples of the effect of numerical approximations (including those in the hardware) are easily found in the area of modelling of nonlinear systems that show a sensitive dependence on initial conditions. Very small differences in the hardware or software design can have large consequences for such systems. One can think here of the use of different floating point precisions (i.e., the number of significant digits used by computer chips) or very small changes in the order of arithmetical operations (which would only produce no differences in outcome if the precision were infinite).

Mistakes, or bugs, in the computer program can remain undetected for a long time, possibly even forever. These bugs cause the computer not to do exactly what it is supposed to do (i.e., numerically solve the intended mathematical equations), which may undermine the theoretical interpretation of a simulation. Given that the material realisation of a simulation is reproducible, the effects of bugs are also reproducible: Each time the model is run, the bug has the same effect. Simulationists should always be aware of this possibility and therefore regularly review their programs. A better strategy is to involve other simulationists in such reviews.

From this discussion on the reproducibility of simulation, we can conclude that with the stability of computer systems, reproducing simulations on the

---

* This is not the case, however, if 'real' stochastic processes like radioactive decay are used to generate 'random numbers' within the program.
† If this second type of reproducibility is attained by a simulationist who wants to transfer a simulation model from one computer to another computer, this does not yet mean that other users necessarily have the skills to use the model. They need sufficient skills with the model to able to actually reproduce simulations.

same computer system is typically unproblematic. Transferring computer models to other computer systems, or building new model implementations, can provide checks on the effects of numerical approximations and mistakes in the original computer programs.

2.3.4 Processed Output Data and Their Interpretation: Visualisation and Understanding

Hacking distinguishes 'marks' as a third category of elements in laboratory practice besides ideas and things. The category of marks is broadly construed to include data and manipulations of data (data assessment, data reduction, and data analysis), as well as interpretations of data. This category of marks is distinguished from the categories of things and ideas because of the distinct role that is played by data in scientific practice. This is also true for simulation practice. Simulations produce a lot of data, which need to be processed. The processing of data produced in simulation practice is similar to that in experimental practice. Often, exactly the same software is used, for instance, for doing statistical analyses on data. At a basic level, the drawing of inferences from the data works quite similarly.

An important element in the practice of processing simulation output in computer simulation is the visualisation of data by means of processing and visualisation software. For example, a visualisation result from my own simulation practice is shown in Figure 2.2. By using advanced visualisation techniques, including animation, simulationists can become familiar with their objects of study in unprecedented ways: They can reach a better

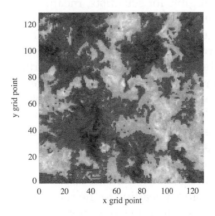

FIGURE 2.2

Horizontal cross section (*xy* plot) of the instantaneous vertical velocity field w at the middle of the convective atmospheric boundary layer as simulated by LES. A linear scale of gray shades is used with a discontinuity at $w = 0$. Light shades correspond to upward velocities. The spatial resolution of the LES is $130 \times 130 \times 66$ grid cells on a $6 \times 6 \times 1.5$ km spatial domain. (From Petersen, A.C. (1999b), Convection and chemistry in the atmospheric boundary layer, PhD dissertation, Utrecht, the Netherlands: Utrecht University, 7.)

understanding of the processes they are studying. In Figure 2.2, for instance, one can *see* the updraughts and downdraughts and their horizontal shapes in the middle of the boundary layer. Pictures of their vertical shapes and 3-D animations also facilitate an intuitive understanding of the behaviour of these plumes. Karim Benammar (1993: 80) writes about the role of visualisation in simulation practices:

> Images are vital for understanding spatial relations between numbers and for discerning patterns that involve more than two numbers. In a graph, only one relationship between numbers is plotted; the advantage of three-dimensional, color graphics is that they make it possible to discern several patterns. The fact that the human brain can process three-dimensional color animations with some facility is a result of both the structure of our image-processing apparatus and our cultural understanding of the representational properties of an image. Our capacity to endow constructed pictures with meaning involves our ability to perceive and process pictures, to focus on specific aspects or generalize from a global perspective, and to see beyond this particular picture.

In the case of the updraught–downdraught model that I developed, diagrammatically pictured in Figure 2.1, I could use the visualisation to check my understanding of the theory that describes the vertical flow within plumes and the exchange of air between plumes (the equations for this can be found in Petersen et al. 1999).

But what do we mean by *understanding* here? As Henk de Regt and Dennis Dieks argue, a phenomenon can be understood if there is a theory of the phenomenon that is intelligible for scientists, meaning that 'they can recognise qualitatively characteristic consequences of [the theory] without performing exact calculations' (de Regt and Dieks 2005: 151). I claim that the equations of fluid dynamics provide such a theory for understanding the phenomenon of turbulent convection in the atmospheric boundary layer. By processing the output data of simulations based on these theories and visualising these data, it is possible to test whether one qualitatively recognises the consequences of the underlying equations.

Indeed, 3-D pictures and animations are ubiquitous in simulation science (see, i.e., the attractive pictures in Kaufmann and Smarr 1993). There is also a danger in their attractiveness: The choices that are made in the processing of the output data (e.g., the projection methods chosen to produce the pictures) are to a large extent arbitrary, leading to the risk of incorrect conclusions (one can see, i.e., side effects of the projection methods instead of real processes). One strategy simulationists use to deal with this is to change the properties of the projection method interactively: If the structure withstands these manipulations, it is most likely not an artefact of the projection method.*

---

* But it can still be an artefact of the simulation model, that is, not represent real processes.

2.4 Plurality of Methodologies for Model Development and Evaluation

A wide variety of methodological approaches exists for performing each of the activities in simulation. I focus here on the activity of (re)formulating conceptual and mathematical models. Simulationists indeed do not follow one dominant methodology when they develop and evaluate their simulation models. The concept of *methodology* is here taken to encompass modelling philosophy, heuristics, and norms for doing 'good' science. The concept of *heuristics* is defined as a set of nonalgorithmic rule-like procedures ('rules of thumb') that can help to achieve certain goals.

Many methodologies can be discerned in simulation practice. I relate some of these methodologies to three general scientific methodologies described in the philosophy of science literature: Karl Popper's philosophy of falsification, Imre Lakatos' 'methodology of scientific research programs', and Leszek Nowak's 'method of idealization and concretization'.[*] My claim is that through the lens of these philosophies some important methodological disagreements in simulation practice can be highlighted. Although most of these philosophies do not provide specific guidance to scientists in terms of how they should develop and evaluate their models, they do lend support to specific heuristical strategies observable in practice.

Many practicing simulationists claim to follow Karl Popper's (1959) philosophy of falsification.[†] Other scientists claim that model development is an 'art' and that there is 'no methodological crank to turn' (Harvey et al. 1997: 7). Popper's basic idea was that scientific progress comes about through testing and falsification. Theories or theoretical systems—and simulation models, for that matter—cannot be verified according to Popper since it can never be known whether they adequately cover all possible cases. The spirit of Popper's philosophy of falsification is aptly summarised by him as follows:

> [The] aim [of the empirical method] is not to save the lives of untenable systems but, on the contrary, to select the one which is by comparison the fittest, by exposing them all to the fiercest struggle for survival. (Popper 1959: 42)

---

[*] Nowak's 'method of idealisation and concretisation' has been taken up by Nancy Cartwright (1989), among others. Note that Cartwright later revised her view on the relation between models and theories (see page 24, second note).

[†] This becomes evident when scientists are asked about their 'philosophy of science'. In many scientific fields, methodological papers have been written by leading scientists who proclaim Popper's philosophy of falsification (for the atmospheric sciences, see, e.g., Randall and Wielicki 1997).

Popper's philosophy does not offer much guidance on what to do after a particular simulation model has been 'falsified'.* What can be distilled from his
philosophy with respect to model testing is the methodological advice not
to focus on verification of simulation models but instead on falsification of
these models. Thus, Oreskes et al. (1994: 641) conclude that 'verification and
validation of numerical models of natural systems is impossible', and they
argue that 'if a model fails to reproduce observed data, then we know the
model is faulty in some way' (Oreskes et al. 1994: 643).†

However, in the practice of simulation, the activities of 'validation' and
'verification' are deemed crucial by most simulationists. Generally, validation is concerned with the establishment of legitimacy (has the *right* model
been built, given the objectives for the model?), and verification is concerned
with establishing truth (is the model *right*)? In operations research, validation
and verification have received somewhat deflated meanings that are not susceptible to Oreskes et al.'s critique of validation (see, e.g., Balci 1994). Model
validation then becomes substantiating that the model, within its domain
of applicability, gives accurate results (with the desired accuracy given by
the study objectives). And model verification is then limited to substantiating that the technical model implementation is an accurate transformation
of the conceptual and mathematical model into a computer program. The
way around the logical problem identified by Popper in his philosophy of
falsification is not to claim universal validity but to limit the model to a particular domain and to accept a particular level of inaccuracy. And, verification can be brought into connection with the second type of reproducibility,
the reproducibility of a simulation under the same theoretical interpretation
(conceptual and mathematical model and result) with a different technical
model implementation. The question is: Have any approximations, assumptions, or mistakes been made in the transformation of the conceptual and
mathematical model to the technical model that jeopardise the accuracy of
the technical model?

Popper's philosophy also entails the following methodological norm:
When trying to come up with an improved theory, do not add ad hoc corrections to the old theory, that is, corrections without separate theoretical
justification (cf. Popper 1959: 42). But, models often *do* contain ad hoc corrections. The methodological norm should therefore be weakened to reflect the
assumption held by many simulationists that the more ad hoc corrections
a model contains the worse it is (see, e.g., Randall and Wielicki 1997). If it is
really found necessary to introduce artificial correction factors, or 'properties of convenience', simulationists should strive to provide an independent

---

* This should come as no surprise since Popper was mainly concerned with the context of
  justification and kept this context distinct from the context of discovery, like most of the
  philosophers of science for most of the 20th century.
† Oreskes et al. do not state that the focus of modelling research should be on falsification.
  According to them, the primary function of models is heuristic. They furthermore conclude
  that 'models can only be evaluated in relative terms' (Oreskes et al. 1994: 641).

justification for these corrections (preferably by deriving them from theory through approximation). This would ensure that the model is based as much as possible on theory instead of letting the model become nagged by auxiliary hypotheses that are not independently justifiable (cf. McMullin 1985: 261). An example from a reflection by David Randall and Bruce Wielicki (1997) on the practice of 'tuning' in simulation may illustrate this point. Tuning consists of adjusting parameters in a model to improve the agreement between the model results and existing measurements. Randall and Wielicki (1997: 404) call tuning 'bad empiricism', and they add that

> although empiricism will always be a necessary part of parameteriza-tion, tuning is not necessary, at least in principle. Genuinely empirical parameters, that is, those that can be measured and are universally applicable, can be set once and for all before a model is run. (Randall and Wielicki 1997: 405)

However, in scientific simulation practice, simulationists hold divergent views of the norm of not adding ad hoc corrections to models. This can be expected from the existence of the third type of simulation identified by Keller (involving models of phenomena for which no general theory exists). To give an example, many complicated simulation models of complex systems such as the environment do contain a large number of ad hoc corrections—often using parameters that are set by tuning—and many simulationists do not find this fact deeply problematic. However, other simulationists judge the models to be relatively unreliable because of the presence of ad hoc corrections and aim at improving the models by removing ad hoc corrections, either by providing theoretical justification for existing corrections (thus removing their ad hoc character) or by replacing ad hoc corrections by changes in models that are considered theoretically justified. For most of these practitioners, *theory* includes empirically established regularities or phenomenological laws. Some scientists have expressed their hope that ultimately all ad hoc corrections currently present in their models can be removed, and that only approximations to empirical laws will remain (e.g., Randall and Wielicki 1997).

The ad hoc corrections that scientists do introduce in their simulation models—against Popper's advice—are usually located in particular parts of the model, parts in which the simulationists are willing to make changes. Mathematical structures and parameters in the models that are already based on well-established empirical laws are typically left unchanged. Examples in the case of 3-D climate models are the conservation laws of momentum and heat (cf. Küppers and Lenhard 2006). This does not entail, however, that the ad hoc corrections are innocent. In the same case of climate models, for instance, changes to the cloud parameterisations that are not yet based on well-established empirical laws may have a huge impact on the results (see Chapter 6).

The philosopher of science Imre Lakatos built his whole methodology of science (his 'methodology of scientific research programs') around the idea that some parts of models are adjusted to reduce the discrepancies between models and data and some parts (the 'hard core') are not. According to Lakatos, a research program

> consists of methodological rules: some tell us what paths of research to avoid (*negative heuristic*), and others what paths to pursue (*positive heuristic*). ... The negative heuristic forbids us to direct the *modus tollens* at [the] 'hard core'. Instead, we must use our ingenuity to articulate or even invent 'auxiliary hypotheses', which form a *protective belt* around this core, and we must redirect the *modus tollens* to *these*. ... The positive heuristic sets out a programme which lists a chain of ever more complicated *models* simulating reality: the scientist's attention is riveted on building his models following instructions which are laid down in the positive part of his programme. (Lakatos 1970: 132, 135)

If we translate Lakatos's view into methodological advice for developing simulation models, it runs like this: Be aware of what constitutes the hard core of your models and refrain from making ad hoc corrections to that hard core.

Indeed, in several scientific simulation practices, Lakatos's distinction between 'hard core' and 'protective belt' seems to hold, which enables one to identify 'families' of models in these practices that share the same hard core but differ in their ad hoc corrections. Also 'lineages' of models can be identified in which old hard cores have been retained within new hard cores (when new scientific research programs branched off from their parent research programs). A case in point is the development of climate models from zero-dimensional (globally averaged) models to 1-D (horizontal dimension from pole to pole), 2-D (either two horizontal dimensions or one horizontal and one vertical dimension), and 3-D models. The 'method of idealisation and concretisation', which was proposed by Leszek Nowak (see, e.g., Nowak 1985: 195–200), identifies stepwise 'concretisations' of 'idealisations'—such as the lineages of models discussed here—as the hallmark of science. Even though Lakatos's and Nowak's philosophies cannot be considered fully adequate descriptions of all scientific practice,* several heuristics found in scientific practice are supported by these philosophies.

The simple—but very influential—methodological advice for developing simulation models that can be inferred from Nowak's methodology is the following: Try to make your model ever more realistic. Models typically do not represent all the details of phenomena (they can be said to contain 'unre-

---

* There are several limitations to the applicability of these philosophies. Firstly, the hard core is never so hard that it forbids all alterations (see the case of Bohr's atomic theory in Radder 1982). Secondly, the method of stepwise concretisation does not capture all important features of scientific theorising or, for that matter, scientific simulation (cf. Kirschenmann 1985: 228–232).

alistic idealisations'). In simulation practice, depending on the purpose of a model, this is not necessarily seen as a problem. Increased realism does not necessarily entail increased accuracy, and decreased realism does not have to result in decreased accuracy. An example of the latter situation is a zero-dimensional climate model that aggregates all surface temperatures into a single zero-dimensional variable of globally averaged surface temperature. Such a mode can give an accurate description of globally averaged quantities, even though it is 'unrealistic' in the sense that the quantities are not homogeneously distributed in reality.

The methodology of concretising a model in a sequence of steps and comparing the results of successive models with the data can be connected with the notion of replicability introduced in Section 2.3.3. In scientific simulation practice, many comparisons are made between the results of models at different levels of concreteness. As soon as the result of a model at a particular level of concreteness is replicated by models at higher levels of concreteness, it can be concluded that the additional details do not make the results more accurate and are not needed for obtaining the results. This can be illustrated by the example of turbulent fluids. The *concreteness* of a model is here defined as the level of aggregation, in terms of both the number of degrees of freedom (or variables) in the model and the number of processes modelled. From comparing the results of more and less-concretised 3-D simulation models of turbulent fluids, it has been found that the behaviour of the fluids at the largest scales in the models is insensitive to the details at the smallest scales. Thus the results pertaining to phenomena at the largest scales can be replicated by using a much less-concretised model. In this case, we can formulate a convenient 'minimal model' that is still sufficiently concrete for adequately simulating some turbulence phenomena (see Goldenfeld and Kadanoff 1999: 87).

Computer simulation is extensively used by scientists studying turbulent fluids. More generally, computer simulation is pervasive in the study of complex nonlinear systems. In developing models, choices must be made with respect to how concrete, or complex, the models should be. The following two competing overarching methodologies can be identified for modelling complex systems:

(M1) to capture as many as possible of the degrees of freedom of the system in a model;

(M2) to keep a model as simple as possible, for instance, by demonstrating that the results of interest can already be replicated by the relatively simple model.[*]

---

[*] Paraphrasing a quotation attributed to Einstein: 'Everything should be made as simple as possible, but no simpler'.

The example of the minimal model of turbulence displays the latter strategy, M2. In geophysical fluid dynamics, there is a tendency to regard M1, the methodology of building models of maximal complexity, as the ideal:

> to the extent that we have good mathematical models of fluid systems, such models usually correspond to essentially nonaggregated representations of the system. (Stevens and Lenschow 2001: 285)

Here, *nonaggregated* models are models at a maximum level of concreteness, that is, with a maximal number of degrees of freedom; such models explicitly resolve large ranges of spatial and temporal scales in modelling the flows (vs. *aggregated models*, which feature a smaller number of degrees of freedom). However, against the general tendency to strive for ever-more-complex models, simple models still have their roles to play in geophysical fluid dynamics. The reasons for this continued role are twofold: (1) With simple models it is easier to grasp what is going on in the model, thus facilitating understanding of a phenomenon by way of the model; and (2) for very large-scale modelling problems, such as those involving Earth's entire system over geological timescales, we cannot do without simple models since the computational resources to run the most complex models for those problems are lacking (Petersen 2004).

2.5 Plurality of Values in Simulation Practice

The choice of methodology by simulationists (e.g., preferring either complex or simple models) depends on the goals of the scientific study involved, the research questions, the required accuracy of the results, the practical limitations encountered when trying to build a maximally detailed model, and so on. For instance, the practical issue of limitations in computing power is a central one in many areas of scientific computer simulation, and it puts upper limits on the level of concreteness in the models. But 'metaphysical' ideas of scientists—their intuitions about the nature of the complexity of the problem—may also influence this choice (cf. de Regt 1996). As Chunglin Kwa (2002) argues, basically two views of 'complexity' coexist in science: a 'romantic' view and a 'baroque' view. The romantic view of complexity relies heavily on the notion of 'holism'; the higher levels in the 'hierarchy of nature' are seen as uniting the heterogeneous items of a lower level into a functional whole. Conversely, the baroque view of complexity pays more attention to the lower-order individuality of the many items making up complexity at the higher level. Such deep philosophical differences partly underlie the choice for either simple or complex models.

There is a plurality of aims and values in simulation practice. In general, there are two modes of dealing with simulation models: treating them as 'objects of study' (noninstrumental mode) or using them as 'instruments' or 'technological artefacts' (e.g., Morrison and Morgan 1999).* The plurality of instrumental values in scientific simulation practice is exemplified by the plurality of functions of simulation. Stephan Hartmann (1996: 84–85) distinguishes five major functions of simulation:

1. Simulation as a *technique*, for investigating the detailed dynamics of a system;
2. Simulation as a *heuristic tool*, for developing hypotheses, models, and theories;
3. Simulation as a *substitute for an experiment*, for performing 'numerical experiments';
4. Simulation as a *tool for experimenters*, for supporting experiments;
5. Simulation as a *pedagogical tool*, for providing understanding of a process.

We have already encountered examples of the first four functions in the present chapter. The differentiation of functions varies between scientific fields. Furthermore, the functions may overlap. I give some examples to illustrate this. In turbulence studies, for instance, physicists who want to examine particular aspects of turbulence dynamics closely often use computer simulation as a technique to do this (Goldenfeld and Kadanoff 1999), not only because simulation is typically less costly than experimentation but also because, for many studies, an experimental approach is hardly or not possible. And in high-energy particle physics, the simulations that are used as tools by experimenters for the design of particle detectors are also used by theorists to study the consequences of their theories (Galison 1996; Merz 1999). Furthermore, many simple simulation models that were developed for any of the first four functions can be used as pedagogical tools, provided that they are made transparent and user friendly enough. Finally, many simulations are used heuristically, while they also perform one or more of the other functions. In fact, Oreskes et al. (1994: 644) claim that *all* simulations should be used heuristically, since 'models are representations, useful for guiding further study but not susceptible to proof', even though that is not what happens in practice. The many functions of simulation in scientific practice led Sergio Sismondo to conclude that 'simulation' is a 'messy category', and

---

* One of the first analyses of the instrumental role played by models in science was a book chapter by the philosopher of science Leo Apostel, which began as follows: 'Scientific research utilises models in many places, as instruments in the service of many different needs' (Apostel 1961: 1). Subsequently, Apostel gave a long list of instrumental functions. For a review of recent literature on models as technological artefacts, see Petersen (2000b).

that 'we should resist the urge to do much epistemological neatening' of it (Sismondo 1999: 258). Still, in this study, the task of doing some epistemological neatening is taken up, in the awareness, however, that complete tidiness is impossible.

With respect to the role of value diversity in science, Helen Longino (1990: 5) argues for science in general

> not only that scientific practices and content on the one hand and social needs and values on the other are in dynamic interaction but that the logical and cognitive structures of scientific inquiry require such interaction.

She distinguishes between 'constitutive values' (epistemic values), which 'are the source of the rules determining what constitutes acceptable scientific practice or scientific method' (Longino 1990: 4), and 'contextual values' (nonepistemic values), which 'belong to the social and cultural environment in which science is done' (Longino 1990: 4). I agree with Longino that both epistemic and nonepistemic values have a legitimate role to play in science.[*]

Since simulationists may hold different epistemic and nonepistemic values, the choices they make in developing and evaluating simulations—as far as these choices are influenced by their values—may be different. The epistemic values of simulationists vary from one to another and from one context to another. For instance, according to some scientists, simulation models serve to generate reliable predictions, while others consider models merely as heuristic tools for theory development. Some scientists aim for models that have a wide range of applicability, while others are satisfied with a more narrowly confined domain covered by the model. And some scientists aim at increasing the 'reality content' of the models, while others adopt more instrumentalist views of what constitutes a good model (focusing more on its usability). Following Kloprogge et al. (2011), who introduced a distinction between the general epistemic values held by individuals (such as the ones just mentioned) and more specific epistemic values that are shared by members of scientific disciplines, I distinguish between 'general epistemic values' and 'discipline-bound epistemic values'. To give an example of the latter, it makes a difference whether you ask a physical scientist or a biologist to develop a model of the climate system; they hold different perspectives on what the important processes and methods are and may each want to deal with those processes in the model and apply those methods with which they are most familiar.

---

[*] According to Longino, the contextual values can affect science through background assumptions. Longino's normative view on the role of contextual values is that they make possible 'criticism from alternative points of view', which 'is required for objectivity' (Longino 1990: 76). She locates the 'objectivity' of science in the deliberative processes used to develop knowledge and, if possible, reach agreement (see Douglas 2004 for a comparison of Longino's notion of 'objectivity' with other notions).

With respect to nonepistemic values, a distinction can be made between 'sociopolitical values' and 'practical values' (Kloprogge et al. 2011). In Chapter 5, I discuss the historical example of ad hoc 'flux adjustment' in coupled climate–ocean models, an assumption that not every climate modelling group was willing to make. By introducing this adjustment, however, it became possible for several climate models to be included in scientific studies of human-induced climate change. The social value of wanting to do policy-relevant science, in addition to personal interests and political conviction, led some modellers to make this unphysical ad hoc correction to their models. Practical values have more to do with practical issues, such as delivering results on time (putting time pressure on the simulation process and forcing simulationists to make simplifying assumptions, for instance), remaining within a given budget, and so on.

2.6 The Practices of Simulation and Experimentation Compared

The interactions between experimental and simulation practices in science are myriad and complex.* On the one hand, simulations can be used to design experiments, and simulations can be used together with experiments to produce 'data'. On the other hand, experiments that help to determine the mathematical structure and parameters in simulation models can be performed. Experiments may be used to produce input data, and experiments can also be used to determine the accuracy of simulations in a certain domain. The analysis of simulation is largely an analysis of hybrid simulation–experimentation practices.

A comparison between the analysis in the present chapter of simulation as a laboratory practice and existing accounts of experimental practice makes it possible to draw some conclusions about similarities and differences between simulation and experimental practice. A first similarity is that both practices involve skills. Hence, both practices are described using expressions such as 'experimenting'; 'playing around'; 'tracking error'; 'dealing with locality, replicability, and stability'; and 'tinkering'.† Some of the early simulation practitioners referred to their simulations as experiments because of their 'experimental' way of working, taken to mean a 'concern with error tracking, locality, replicability, and stability' (Galison 1996: 142). While the concept of tinkering certainly captures an important aspect of simulation practice, this similarity between experimental and simulation

---

* The relationship with observational practices is left implicit in this section.
† Theoretical science can also be analysed in terms of practical work that includes tinkering; in that case, it concerns tinkering with equations (Merz and Knorr Cetina 1997).

practice is somewhat superficial and does not distinguish the two practices from other, nonscientific practices. A second similarity between simulation and experimental practices is that both involve models of the outside world, and the results obtained in the laboratory are extrapolated to this outside world.* And a third similarity is that at the practical level we find thing elements within simulation practice that have a function similar to the thing elements in experimental practice.

The first—and main—difference between simulation and experimentation is that the things manipulated in simulation practice are mathematical models materialised in computer programs and not material models, as in experimental practice. The representational relationship between a mathematical model and reality is different in kind from the representational relationship between a material model and reality. A second difference is that in simulation, no background theory or modelling of the apparatus are needed since the computer hardware may in practically all cases be expected to run the software in the way it is engineered to run it. And a third difference is that the norm of the reproducibility of the material realisation is generally more difficult to meet in experimental practice than in simulation practice.

Contrary to my use of the term *experiment* here, Mary Morgan (2003) does not reserve the term for material experiments only. She also discusses 'mathematical model experiments', in which, by way of deduction, the consequences of various interventions in the model are calculated. And she identifies simulations as belonging to an in-between category of 'hybrid experiments'. On the one hand, in simulations the computer is configured 'to produce results for the particular case used, not to deduce or derive general solutions' (Morgan 2003: 222). From this Morgan concludes that the process of demonstration resembles material experiments more closely than mathematical experiments. On the other hand, simulations rely on mathematics. Subsequently, she argues that some simulations look more like material experiments than others (based on the verisimilitude of the input data to reality). As I discussed in Section 2.3.2, this proximity should only be interpreted as closeness of the mathematical input of the simulation to the material input of the material experiment. I conclude that the two types of simulation distinguished by Morgan, namely, 'virtually experiments' and 'virtual experiments', are much closer to each other in kind than their labels suggest. They are both fully mathematical. To stress the difference between mathematical and material manipulations, I refrain from associating the term *experiment* with simulation.

---

* In experimental practices, material models of the outside world are used, and the theoretical interpretation of the phenomena is extrapolated to the world outside the laboratory.

2.7 Conclusion

I have proposed that the activities of simulationists can be grouped conceptually into four main types, and that four epistemologically distinct elements can be identified in simulation practice: (1) the conceptual and mathematical model; (2) model inputs; (3) the technical model implementation; and (4) processed output data and their interpretation. These elements can be connected to the taxonomy of things, ideas, and marks that Hacking developed for the experimental laboratory sciences. I have argued that simulation science can itself be regarded as a laboratory science. My claim is that simulation practice can be fruitfully analysed from the perspective of the laboratory sciences. However, even though many analogies exist with experimental practice, there is a fundamental difference between mathematical and material manipulations. Therefore, a conceptual distinction between simulation and experimental practice should be maintained.

In the discussion of the four main elements of simulation practice, four philosophical questions were addressed. First, I argued that the distinction between general theory and models should be considered to be a relative one. Second, on the proximity of simulation and material experiments, I concluded that by using real-world input, a simulation can be used as part of a measurement apparatus in experiment or observation, but that the extent to which the outcomes are reliable depends not only on the input data, but also on the reliability of the conceptual and mathematical model. Third, I argued that (1) reproducing simulations on the same computer system is typically unproblematic and (2) transferring computer models to other computer systems, or building new model implementations, is more difficult but can provide checks on the effects of numerical approximations and mistakes in the original computer programs. And fourth, I showed that by using advanced visualisation techniques, including animation, simulationists can gain a better understanding of the processes they are studying.

By relating three general scientific methodologies (those of Popper, Lakatos, and Nowak) to the practice of simulation, it was found that these may be used to generate heuristics that can actually be identified in examples of simulation practice. However, there is no consensus on the application of these heuristics. This situation reflects the plurality of aims and values in simulation practice.

3

A Typology of Uncertainty in Scientific Simulation

3.1 Introduction

Like the term *error*, *uncertainty* is imbued with many meanings (Kirschenmann 2001). The term may refer to a variety of different things; for instance, people refer to nature, propositions, models, practices, or the future as 'uncertain'. In many of these cases, the term is used in a loose manner. In some formal definitions of the term *uncertainty* as 'absence of certainty' or 'lack of knowledge', uncertainty is taken to refer to our state of knowledge (that is, propositions or models *about* nature, practices, and their future).* A typical example of a definition of uncertainty is the following:

> Uncertainty can be defined as a lack of precise knowledge as to what the truth is, whether qualitative or quantitative. (National Research Council 1996a: 161)

According to this definition, uncertainty is more or less the same as inaccuracy. Funtowicz and Ravetz (1990) provide a broader characterisation of uncertainty. They distinguish between two main dimensions of uncertainty. Their first main dimension is the *source* dimension, which I call the 'location' dimension in this discussion. Their second main dimension is the *sort* dimension:

> Classification by sources is normally done by experts in a field when they try to comprehend the uncertainties affecting their particular practice. But for a general study of uncertainty, we have to examine its sorts. (Funtowicz and Ravetz 1990: 22–23)

Along the sort dimension, Funtowicz and Ravetz list three types of uncertainty: 'inexactness' (imprecision, usually expressed by a spread); 'unreliability' (inaccuracy, usually expressed by a statistical confidence

---

* Note that defining uncertainty as lack of knowledge does not imply that the lacking knowledge can be gained in principle. See the discussion on ontic uncertainty in this chapter.

level); and 'border with ignorance' (not expressed statistically). Furthermore, they elaborate the concept of 'pedigree'. The pedigree of a particular piece of information conveys an evaluative account of the production process of that information.

In this chapter, I argue that Funtowicz and Ravetz's (1990) characterisation of uncertainty is not yet refined enough since there are more sorts that should be distinguished, and that these sorts should be considered as separate dimensions that can be used in parallel to characterise a particular source of uncertainty. Therefore, I propose the following six-dimensional typology of uncertainty:

1. Location of uncertainty
2. Nature of uncertainty
3. Range of uncertainty
4. Recognised ignorance
5. Methodological unreliability
6. Value diversity

This typology is presented graphically in Figure 3.1.* The pedigree of information is represented in this typology by the dimensions of methodological unreliability and value diversity. Funtowicz and Ravetz's notions of inexactness and unreliability are captured by the dimension of range of uncertainty.

In Sections 3.2–3.7, each of these dimensions is introduced consecutively. It must be stressed at the outset that no typology of uncertainty exists that includes all of its meanings in a way that is clear, simple, and adequate for each potential use of such a typology. My claim with respect to the typology given in Figure 3.1 is that it offers a philosophically meaningful insight into the different types of uncertainty that play a role in simulation practice. As such it can also be of use to simulationists who want to assess the uncertainties in their simulations.† Note that although the typology is graphically represented by a two-dimensional matrix, the typology is really six dimensional, and each column plays a role in characterising a particular source

---

* This typology is a synthesis of two lines of typologies, one based on the dimensions *inexactness, unreliability,* and *recognised ignorance* (e.g., Funtowicz and Ravetz 1990; van der Sluijs 1995, 1997) and one based on the dimensions *nature of uncertainty* and *level of uncertainty* (Walker et al. 2003; Janssen et al. 2003; van der Sluijs et al. 2008). I have kept recognised ignorance as a separate dimension—thus taking it out of the 'level' of uncertainty dimension—and have relabelled the remaining cluster of 'statistical' uncertainty and 'scenario' uncertainty as the dimension *range of uncertainty.* In addition, the dimension *value-ladenness of choices* was taken from Janssen et al. (2003).

† How that could be done is elaborated in Chapter 4.

Sorts of Uncertainty

Uncertainty Matrix — Location/Source of Uncertainty →	Nature of Uncertainty		Range of Uncertainty (Inexactness/Imprecision or Unreliability$_1$/Inaccuracy)		Recognised Ignorance	Methodological Unreliability (Unreliability$_2$)	Value Diversity
	Epistemic Uncertainty	Ontic Uncertainty/ Indeterminacy	Statistical Uncertainty (Range + Chance)	Scenario Uncertainty (Range of 'What-If' Options)		• Theoretical Basis • Empirical Basis • Comparison with Other Simulations • Peer Consensus	• General Epistemic • Discipline-Bound Epistemic • Sociopolitical • Practical
Conceptual model							
Mathematical model — Model structure							
Mathematical model — Model parameters							
Model inputs (input data, input scenarios)							
Technical model implementation (software and hardware implementation)							
Processed output data and their interpretation							

FIGURE 3.1

Typology of uncertainty in simulation.

of uncertainty (see, i.e., Figure 6.4, discussed in Chapter 6).* In Section 3.8, I discuss the differences and similarities between uncertainties in simulation and experimentation.

3.2 Locations of Simulation Uncertainty

The daily practice of simulationists is full of uncertainties of many types: They are faced with uncertainties in their models and input data; they may not have sufficient skills to perform a simulation; their organisational politics may be precarious, and so on. Following the argument of Chapter 2, all uncertainties in simulation are ultimately reflected in, or located in, the four main elements of simulation practice: the conceptual and mathematical model; the model inputs; the technical model implementation; or the processed output data and their interpretation. The *location* of simulation uncertainty indicates where the uncertainty manifests itself among the main elements of simulation practice.

3.3 The Nature of Simulation Uncertainty

The 'nature' of uncertainty expresses whether uncertainty is primarily a consequence of the incompleteness and fallibility of knowledge (*epistemic uncertainty*) or whether it is primarily due to the intrinsic indeterminate or variable character of the system under study (*ontic uncertainty*). Given the definition of uncertainty as 'lack of knowledge', ontic uncertainty implies a lack of knowledge arising from the specific ontic character of the object of knowledge. The basic distinction between two natures of uncertainty underlies many binary uncertainty typologies, such as 'aleatory uncertainty' (due to chance) versus 'epistemic uncertainty' (National Research Council 1996b: 107) or 'variability' versus 'lack of knowledge'[†] (e.g., van Asselt 2000: 85). Although the distinction is analytically sound, in scientific practice mixtures

---

* Indeed, many different combinations of characterisations along all six dimensions are possible. Even though the last two dimensions—methodological unreliability and value diversity—will typically be an expression of epistemic uncertainty, one can have methodological unreliability and value diversity in estimates of ontic uncertainty (see Section 3.3). Therefore, I find it useful to keep the last two dimensions distinct and independent from the nature dimension.

† Note that although van Asselt first distinguishes between variability and lack of knowledge as 'the two major sources of uncertainty' (van Asselt 2000: 85), she subsequently indicates that 'lack of knowledge is partly a result of variability' (van Asselt 2000: 86). The latter statement is consistent with my use of the phrase *lack of knowledge*.

of ontic and epistemic uncertainty are often encountered (cf. Kirschenmann 2001: 4). For example, on the one hand, the fact that the weather is unpredictable stems from a fundamental property of chaotic systems (their 'sensitive dependence on initial conditions'). On the other hand, this unpredictability results from our limited knowledge of the initial state, the unreliability of numerical weather prediction models, and other scientific limitations. There is still some room for extending the prediction horizon, the maximum time span over which one can reliably predict ahead. The reason why the distinction between ontic and epistemic uncertainty is useful is that both categories entail different conclusions with respect to the reducibility of uncertainty. Many ontic uncertainties are irreducible, while many epistemic uncertainties are reducible, although the match between the two distinctions (ontic–epistemic and irreducible–reducible) is not perfect.

The meaning of ontic uncertainty in simulation-laboratory practice deserves some further explanation. If the location of ontic uncertainty is the conceptual and mathematical model, then the indeterminate or chaotic character of the system is embedded in the model. In the ideal case in which we judge the model to be perfectly reliable, there is no epistemic uncertainty about the model. However, in practice, models are never perfectly reliable, and we are always faced with ontic uncertainty and epistemic uncertainty, including epistemic uncertainty about ontic uncertainty. Take, for instance, models of the climate system. Within definite bounds, the globally averaged surface temperature on Earth fluctuates in an unpredictable manner due to the natural variability of the climate system. This natural variability consists of two components: an 'internal variability' of the climate system (manifested, e.g., in the El Niño phenomenon) and an 'external natural variability' (related to volcanic eruptions, for instance). To determine whether, statistically speaking, the warming observed since the middle of the 20th century is due to this 'internal variability' of the climate system or to some other causes, such as 'external natural variability' (volcano eruptions, for instance) or human influences, we need to establish the magnitude of the internal variability of the climate system. Determining the ontic uncertainty that reflects this internal variability requires the use of complex climate models; we cannot determine it directly from observations. The models are not perfectly reliable, however, and thus there is epistemic uncertainty with respect to the estimated range of the ontic uncertainty.

3.4 The Range of Simulation Uncertainty

A scientific claim based on results from computer simulation may express a *range of uncertainty*. This range may in turn derive from uncertainty sources at the different locations distinguished in Figure 3.1. In science, uncertainty ranges come in two types: statistical uncertainty and scenario uncertainty ranges. A *statistical uncertainty* range can be given for the uncertainties that can be adequately expressed in statistical terms, for example as a range with associated probability (e.g., uncertainty reflected in modelled variability within a system). Both objective and subjective probabilities can be used. In the natural sciences, scientists generally refer to this category of uncertainty, thereby often implicitly assuming that the model relations involved offer adequate descriptions of the real system under study, and that the (calibration) data employed are representative of the situation under study.

However, 'deeper' forms of uncertainty are often at play. These cannot be expressed statistically but can sometimes be expressed by a range. Such a range is then called a *scenario uncertainty* range. Scenario uncertainties cannot be adequately depicted in terms of chances or probabilities but can only be specified in terms of (a range of) plausible events. For such uncertainties to specify a degree of probability or belief is meaningless since the mechanisms that lead to the events are not sufficiently known. Scenario uncertainties are often construed in terms of 'what-if' statements. In principle, it is possible that uncertainties first expressed as scenario uncertainties later switch to the category of statistical uncertainty if more becomes known about the system processes.[*] This is what has happened with simulation-based estimates of the sensitivity of climate to greenhouse gas increases, the 'climate sensitivity'.[†] The range of climate sensitivities used in the Intergovernmental Panel on Climate Change (IPCC) report was 1.7–4.2°C (IPCC 2001: 13). No statistical characterisation was given of this range. The range was based on the selection of a set of models with different climate sensitivities (with the following reasoning: What is the climate sensitivity if model 1 is true? etc.). This range was still close to the commonly accepted range of 1.5–4.5°C, which has also been characterised statistically in the past.[‡] After the IPCC (2001) report had been published, articles increasingly appeared in the literature that included statistical characterisations, through probability density functions, of the climate sensitivity. In the subsequent assessment report, the IPCC actually decided to move the climate sensitivity into the statistical uncertainty category: 'It is *likely* [defined as a judgmental estimate of a 66–90% chance] to be in the range 2°C to 4.5°C' (IPCC 2007a: 12).

The measure of the spread of both types of uncertainty range is either *inexactness*,[§] also called 'imprecision', which gradually moves from exact

---

[*] The switch can work the other way around (from statistical uncertainty to scenario uncertainty) if one later realises that too little is known.

[†] The *climate sensitivity* is defined as the equilibrium global surface temperature increase for a doubling of the CO_2 concentration.

[‡] See the work of van der Sluijs et al. (1998) for an overview of the ranges that have been given and their meaning. They argue that the range of climate sensitivity has acted as an 'anchoring device', with the range receiving a different meaning with virtually every new assessment of climate change (see Chapter 4 for further discussion of this topic).

[§] The term *exactness* (or *inexactness*) has many meanings (Kirschenmann 1982). I propose to use this one in accordance with Funtowicz and Ravetz (1990).

(small range) to inexact (large range), or *unreliability₁* (defined further in this section), also called 'inaccuracy', which gradually moves from accurate (small range) to inaccurate (large range). While Funtowicz and Ravetz consider a 'spread' to derive typically from 'random error' (e.g., Funtowicz and Ravetz 1990: 23, 70), the interpretation of 'range' in the typology proposed here is much broader and encompasses all types of statistical and scenario uncertainty ranges. Ranges of uncertainty can derive from *any* source of uncertainty in scientific practice, including model structure. Thus, in principle, 'systematic error' can be represented in the typology as statistical uncertainty arising from the model structure.

The statistical uncertainty range can be qualified by information about the statistical reliability (*reliability₁*). The term *(un)reliability* was frequently used in the preceding text, without explicit mention of what precisely was meant. One has to ask the question: What is reliable for what purpose? In this study, I use two notions of 'reliability', denoted as reliability₁ and reliability₂. For scientific simulation laboratory practice, I define *reliability₁* as follows: The reliability₁ of a simulation is the extent to which a simulation, or combination of simulations, yields accurate results in a given domain. It is important here to distinguish between 'accuracy' and 'precision' (see Hon 1989: 474). Accuracy refers to the closeness of the simulation result to the 'true' value of the sought physical quantity, whereas precision indicates the closeness with which the simulation results agree with one another, independently of their relations to the true value. 'Accuracy thus implies precision but the converse is not necessarily true' (Hon 1989: 474). Traditionally, the distinction between 'systematic' and 'random' error is taken to correspond with the distinction between accuracy and precision (Hon 1989: 474). Since systematic and random error are solely statistical notions, I propose to dissociate these two dichotomies from each other, so that all sources of error may be assessed in terms of their impact on the accuracy and the precision of the results.[*]

The definition of reliability₁ refers to a whole simulation that delivers results, not to the elements of scientific practice such as the mathematical model. The term *reliability* is also applied to such elements of scientific practice. Given the definition of the reliability₁ of a whole simulation, it seems most natural to define the reliability₁ of the elements of scientific practice in

---

[*] The theory of statistical error analysis, as is developed for instance by Deborah Mayo (1996) for experimentation, is too limited for a proper analysis of uncertainty in science. Both in the experimental and the simulation laboratory, it is often not a straightforward exercise to determine the reliability₁ of a simulation; there are many different factors involved that could cause error. The standard view of error in scientific textbooks holds that, 'apart from random errors, all experimental errors can be eliminated', a view that is grossly mistaken, however, since the complexity of actual reality typically prevents systematic errors from being reducible to zero (Hon 1989: 476). The distinction between systematic and random errors is only mathematically based and has only limited value for actually determining error in scientific practice. I therefore agree with Hon (2003: 191–193) that the theory behind the concepts of random and systematic errors is purely statistical and not related to the locations and other dimensions of uncertainty.

terms of their effects on the accuracy of the final results. Since errors may cancel each other out, we can never infer the reliability$_1$ of the elements of the simulation from the reliability$_1$ of the simulation as a whole ('from the top down'), given that we were able to establish the latter.* The question of whether a particular element is reliable$_1$ thus depends on the circumstances under which the element is used and needs to be established 'from the bottom up'. There is a whole specialised field of sensitivity analysis and a suite of methods that can be employed in

> the study of how the variation in the output of a model (numerical or otherwise) can be apportioned, qualitatively or quantitatively, to different sources of variation, and of how the given model depends upon the information fed into it. (Saltelli 2000: 3)

If one also determines what the uncertainties in the model input, model structure and model parameters are, the reliability$_1$ of the whole simulation can be estimated through mathematical techniques of uncertainty analysis. In practice, however, not all information needed to determine the reliability$_1$ of the different elements is available, and one therefore will often have recourse to qualitative methods of determining reliability (see Section 3.6).

Thus statistical uncertainty ranges can be determined either from comparing the simulation results with measurements (provided that accurate and sufficient measurements are available) or from uncertainty analysis (provided that the accuracy of the different elements in simulation is known). Scenario uncertainty ranges (based on what-if questions) are generally easier to construct by varying simulation elements.

3.5 Recognised Ignorance in Simulation

Recognised ignorance about a phenomenon we are interested in concerns those uncertainties that we realise—in one way or another—are present, but for which we cannot establish any useful estimate, for example, due to limits of predictability and knowability ('chaos') or due to unknown processes. *Unrecognised ignorance* does not count as uncertainty since it concerns 'pure ignorance' about which we cannot say anything knowledgeable. Experts making a claim may acknowledge that they are ignorant about particular locations of uncertainty. For instance, in the case of a complex system, it can be acknowledged that the system may show surprises about which one

---

* A way out could be to establish the reliability of the elements by applying them under different circumstances, that is, in different simulations, and trying to find out to what extent they affect the reliability of the different simulations. The success of this approach is limited by the fact that multiple errors could exist simultaneously.

only has qualitative knowledge and that are not reflected in the model structure used. Another example relates to the adequacy of the software implementation. Good simulation practitioners are aware of the fact that bugs can easily slip into their code and that—until they find them—they are ignorant of their precise location in the program.

A measure of recognised ignorance is the *openness* of a claim: the presence of an explicit reflection within the claim of the fact that one is aware of one's ignorance about particular sources of uncertainty. Closed claims, which do not offer such reflection, convey an appearance of certainty about the uncertainties reflected in the claim. The openness of a claim can be expressed in a general manner, such as in 'surprises are not excluded', or with reference to particular uncertainties (without giving uncertainty ranges since these are acknowledged to be unknown). Scientists can also give their subjective probability for a claim, representing their estimated chance that the claim is true. Provided that they indicate that their estimate for the probability is subjective, they are then explicitly allowing for the possibility that their probabilistic claim is dependent on expert judgement and may actually turn out to be false.*

As will become evident in Section II of this study, the practice of climate simulation is full of recognised ignorance. To give an example, on 26 May 2006 the journal *Geophysical Research Letters* published two articles that both claim that the positive feedback between higher temperatures and higher CO_2 concentration is not adequately represented in most climate models (Scheffer et al. 2006; Torn and Harte 2006). From an analysis of historical (proxy) data for several hundred thousand years, they claim that the effect is much stronger than has previously been estimated. Scheffer et al. acknowledge their ignorance about which processes actually cause the positive feedback. They write:

> The main merit of our approach as we see it, is that it allows for an estimate of the potential boost in global warming by century-scale feedbacks which is quite independent from that provided by coupled CO_2–climate models that explicitly simulate a suite of mechanisms. Like our approach these models have considerable uncertainty. Not only are the quantitative representations of the mechanisms in the models uncertain, there is also always an uncertainty related to the fact that we are not sure whether all important mechanisms have been accounted for in the models. (Scheffer et al. 2006: 4)

Thus while Scheffer et al. claim that their results are independent of the most relevant underlying mechanisms, they clearly recognise the ignorance with respect to these mechanisms, which may not yet have been included in present climate-simulation models.

---

* To be fair, also estimates of objective probabilities may turn out to be false. This is why the subject of the reliability$_2$ of a claim, treated in Section 3.6, is so important.

3.6 The Methodological Unreliability of Simulation

As was observed in Section 3.4, a major limitation of the statistical definition of reliability is that it is often not possible to establish the accuracy of
the results of a simulation or to assess quantitatively the impacts of different
sources of uncertainty. In those cases, one may have recourse to *qualitative*
judgements of the relevant procedures instead. Scientists can, for instance,
judge the methodological rigour of the scientific procedure followed—the
methodological rigour may then be regarded as an alternative for accuracy. Let me therefore also give a methodological definition of reliability,
denoted by reliability$_2$: The *reliability$_2$* of a simulation is the extent to which
the simulation has methodological quality. The methodological quality of
a simulation derives from the methodological quality of the different elements in simulation practice. The methodological quality of a simulation, for
example, depends not only on how adequately the theoretical understanding of the phenomena of interest is reflected in the model structure but also,
for instance, on the empirical basis of the model: the numerical algorithms,
the procedures used for implementing the model in software, the statistical
analysis of the output data, and so on.

While the range of uncertainty is a *quantitative* dimension of uncertainty,
the other five dimensions (including location) are *qualitative*. Since methodological quality is a qualitative dimension and the (variable) judgement and
best practice of the scientific community provides a reference, determining
the methodological quality of a claim also is not usually a straightforward
affair. It depends, for instance, on how broadly one construes the relevant
scientific community. The broader the community, the more likely it is that
the different epistemic values held by different groups of experts could influence the assessment of methodological quality. Criteria such as (1) theoretical basis, (2) empirical basis, (3) comparison with other simulations, and (4)
acceptance/support within and outside the direct peer community can be
used for assessing and expressing the level of reliability$_2$. Each of these four
criteria are now dealt with in turn.

3.6.1 The Theoretical Basis of Simulations

As was discussed in Section 2.3.1, the conceptual and mathematical models
used in simulation vary in the extent to which they are derived from general
theory by way of approximation, that is, vary in their theoretical quality.
My claim is that if the only relevant methodological difference between two
models is their level of theoretical quality, then the model with a higher theoretical quality should be considered more plausible.

But can we also expect a higher reliability$_1$ for simulations with a stronger theoretical basis if all other things are equal? Nancy Cartwright
(1983) argues that fundamental laws themselves do not generate accurate

predictions, but that they need to be translated into models that can do the job for them:

> [F]undamental equations are meant to explain, and paradoxically enough the cost of explanatory power is descriptive adequacy. (Cartwright 1983: 3)

This is a consequence of multicausality. Each cause by itself cannot contribute to descriptive adequacy, but appropriately taken together with other causes, in a model, the model can be descriptively adequate. She proposes a *'simulacrum* account of explanation':

> The route from theory to reality is from theory to model and then from model to phenomenological law. The phenomenological laws are indeed true of the objects in reality—or might be; but the fundamental laws are true only of the objects in the model. (Cartwright 1983: 4)

Cartwright's (1983) view pertains to Keller's second type of simulation: simulation involving conceptual approximations to general theory. I agree with Cartwright's argument that the accuracy of the model improves if there are underlying fundamental laws that are true of the objects in the model.[*]

This can be illustrated thus: Let us assume that we have two models that accurately describe the phenomena in a particular domain. One model is derived by way of approximation from a fundamental law. This theory has a scope that is larger than the model domain.[†] The other model is not derived from theory. The question that then arises is whether the fact that the first model is derived from theory positively contributes to the reliability$_1$ of a simulation outside of the original domain to which the model was applied (but within the scope of the fundamental law). As long as we cannot strictly establish the reliability$_1$ of the model, we are left with establishing the reliability$_2$, which in this case indicates a relatively high theoretical quality for the first model as compared with the second model. We may subsequently hypothesise that the reliability$_1$ of the first model outside the original domain is higher than that of the second model.

From this argument, I conclude that it is important both to determine the extent to which simulation models are derived from general theory and furthermore to determine the scope of the general theory.

---

[*] Note, however, that Cartwright has changed her view (see page 24, second note).
[†] This follows from the constraint structure for theories, which entails that they are more universal than models (Weinert 1999).

3.6.2 The Empirical Basis of Simulation

The methodological criterion for the empirical basis of a simulation should distinguish between simulations that are based on or have been tested against sparse observations or unreliable$_2$ experiments, on the one hand (low quality), and simulations that have been thoroughly built on empirical input (high quality) on the other. But the qualitative assessment of the fit between the simulation and the system of study also belongs to the assessment of the empirical basis.* Mary Morgan (2003: 233) emphasises that 'we need to look carefully at where and how much materiality is involved [in simulation], and where it is located, before we can say much more about its validity'. Thus, the stronger the empirical basis of a model, the more reliable$_2$ it is. Morgan, however, only considers the model inputs, and thus she neglects the assessment of the empirical basis of the simulation model itself.

There are several ways to compare the simulation model with experiments and observation. Experiments may be specifically designed or observations may be specifically planned for evaluating models (see, e.g., Siekmann 1998). Simulation models that have been evaluated in this way have a stronger empirical basis.

By cleverly choosing the measurements, a group of atmospheric scientists have developed a methodology for assessing their models, adherence to which would lead to a stronger empirical basis of models. Specifically, they proposed to do experiments and to make observations to evaluate large-eddy simulation (LES) models (Stevens and Lenschow 2001). In the past, LES models have been tested against existing atmospheric observations. Furthermore, they have been compared with 'mimicking experiments' (material scale models of the atmosphere) involving water tanks. The comparison suffers from a reliability$_2$ problem related to the water tanks. However, the water tanks need not adequately represent the atmosphere since the 'turbulence intensity' (a measure of how turbulent a flow is) in the tank is much smaller than the turbulence intensity in the atmosphere. Since the group of atmospheric scientists was not satisfied with the way experiments and observations had been used in the past to evaluate LES models, they proposed three guidelines for evaluating models based on experiments or observations.

Since these guidelines have a broader applicability, not restricted to LES models alone, they are given a more general formulation here. The first guideline is the following: Design field observations or laboratory experiments specifically to test details of your model. The fact that often only existing data are used limits the strength and precision of statements that can be made about the reliability$_1$ of parts of a model and about the areas where improvement is needed. The second guideline is as follows: Be clear on what

---

* The quantitative assessment of the comparison between simulation and experiment or observation may lead to estimates of the range of uncertainty (accuracy).

constitutes a significant difference between model and observations and take the accuracy of the measurements into account (also valid when using existing data). This must prevent scientists from either erroneously accepting or rejecting the model as accurate for the quantities of interest. Finally, the third guideline is to base a test of your model particularly on quantities that constitute relevant outcomes of your model. If, for example, LES models were evaluated on the basis of quantities that are also generated by much simpler models, the potential 'added value' of the LES models—the possibility of providing detailed information about turbulent flows, which constitutes the main reason why the models are used—would not be assessed.

3.6.3 Agreement of Simulations among Each Other

If the result of a simulation is replicated by another simulation, then the original simulation can be considered to have become more reliable$_2$, depending on how the other simulation was done and how reliable$_2$ the other simulation is. One strategy simulationists may follow to increase the methodological strength of their simulations is to relate the results of their simulations to the results of other simulations of similar processes. The third type of reproducibility (discussed in Section 2.3.3), the reproducibility of the result of a simulation, or replicability, is a very important type of reproducibility in simulation practice.[*]

3.6.4 Peer Consensus on the Results of Simulations

The consensus about the results of a simulation can be determined via peer review, which is a mechanism for quality control in science. The role of peer review in determining the reliability$_2$ of one's conclusions is well established. Popper, for instance, emphasises the importance of peer review. His philosophy of falsification, introduced in Chapter 2, features a sensitivity to the fact that falsification is not a matter of simply applying an algorithm. Popper admits that in scientific practice, while the idea of falsification is based on a simple logical scheme, it needs judgment by experts.[†] For simulation models, this means that a conclusive disproof of a simulation model is impossible since either the experimental results with which it is compared may be

---

[*] It may be the case that within a group of models all models agree, that is, replicate each others' result, but that they are all unreliable$_1$, for instance, if they are all built on the same unreliable$_1$ principles. To guard against those situations, comparison of model results with empirical data remains crucial.

[†] Popper writes: 'A theory is a tool which we test by applying it, and which we judge as to its fitness by the results of its application' (1959: 108). It is the notion of 'judgment' that I wish to emphasise here. According to Popper, science involves the application of 'methodological rules', which 'are very different from the rules usually called "logical". Although logic may perhaps set up criteria for deciding whether a statement is testable, it certainly is not concerned with the question whether anyone exerts himself to test it' (Popper 1959: 54).

judged to be unreliable₁ or the discrepancies may be considered unimportant. In the latter case, the discrepancies may be assumed to disappear with new versions of the simulation model, for instance. Popper was aware of the fact that it is our decisions that settle the fate of theories—or models, for that matter—and that 'this choice is in part determined by considerations of utility' (Popper 1959: 108). 'From a rational point of view', we should never definitively embrace any theory, but we should 'prefer as basis for action' the best-tested theory (Popper 1979: 22). The best-tested theory is the one which, 'in the light of our *critical discussion*, appears to be the best so far' (Popper 1979: 22). Critical discussion, not only with respect to science but also more generally, thus represents the core of Popper's philosophy of critical rationalism. The methodological advice that can be derived from his critical rationalist position is the following: Make active use of peer review and critical discussion of your model.

3.7 Value Diversity in Simulation Practice

In Chapter 2, I introduced two types of epistemic values (general epistemic values and discipline-bound epistemic values) and two types of nonepistemic values (sociopolitical values and practical values) that play a role in simulation practice. The sixth dimension of uncertainty is the *value diversity* reflected in the different assumptions made in simulations. Simulationists often have considerable freedom in making choices with respect to the conceptual and mathematical model; the model inputs; the technical model; and the processing of output data and their interpretation. These choices are made either implicitly or explicitly. They have a subjective component and may be influenced by epistemic and nonepistemic values held by the simulationist. The choices thus have a potential to be value laden. If the value-ladenness is indeed high for specific elements of the assessment and the results are significantly influenced by the value-laden choices made, then the simulation results are also value laden. This is, for instance, the case for climate models that employ the unphysical 'flux adjustment', which has a significant influence on the results (see Chapter 5). The presence of value-laden assumptions is then also reflected in the other dimensions of uncertainty (e.g., scenario uncertainty and unreliability₂ with respect to the assumed model structure in models that include flux adjustment).

In some cases, the scope and robustness of the conclusions of the study may be limited, that is, 'biased',* by this value-ladenness. Although smaller

---

* *Bias* is used here to indicate a systematic influence of values and should not be interpreted as a pejorative term.

groups of experts may make value-laden assumptions more easily without questioning them, whole disciplines may also have particular biases.

___

3.8 The Uncertainties of Simulation and Experimentation Compared

In Chapter 2, it was observed that both simulation and experimental practices involve models of the outside world (mathematical models in simulation practice and material models with theoretical interpretations of the phenomena in experimental practice). Thus both practices are confronted with the various types of uncertainties in models. More generally, many of the sources of uncertainty located in Hacking's category of 'ideas' are similar for experiments and simulations. In both cases, the theoretical concepts and models used may be more or less reliable$_1$, and errors may be made in the interpretation of the outcomes (see my discussion of 'marks' in Section 2.3.4), leading to erroneous theoretical conclusions (cf. Hon 1989). This similarity is often not acknowledged in controversies on the reliability of experiments versus the reliability of simulation, such as the ones Chapter 1 exemplified.* Often these models remain hidden from view, which can then obscure the fact that the results of measurements are sensitive to modelling assumptions.

The main difference between simulation and experimentation is that there is no physical nature present that is under study in simulation laboratories, while there is in experimental ones. We can conclude that this gives rise to an additional uncertainty in experimentation: Due to disturbing processes, the target may behave differently from the way expected. While in simulations the consequences of our mathematical models may be unexpected and 'surprise' us, we may be 'confounded' (Morgan 2003: 221) by the behaviour of the natural processes that differs from what we expected. According to Morgan, 'this suggests that material experiments have a potentially greater epistemological power than nonmaterial ones' (Morgan 2003: 221). In simulation, we may erroneously decide not to include particular processes in a model and not be immediately confronted with the consequences of this omission. In experimentation, however, such processes can directly interfere with the experiment. Still, I maintain that the uncertainties in experimentation are not necessarily smaller than the

___

* The same is true for observations. Some observational measurements, for instance, strongly depend on models, but they are often presented as unproblematic measurements (e.g., satellite measurements). It may also happen that the comparison of a simulation model result with an observational measurement may not lead to unequivocal conclusions if the observation is dependent on another simulation model (e.g., when comparing historic runs of climate models with weather measurements derived from numerical weather prediction models).

uncertainties in simulation, and that in particular cases simulations are more reliable than available experiments.

3.9 Conclusion

By extending Funtowicz and Ravetz's (1990) typology of uncertainty and combining it with the elements of simulation practice identified in Chapter 2, I have arrived at a more complete account of the types of uncertainty that play a role in simulation practice. The notion of uncertainty developed here is broader than statistics. While statistical uncertainty—which is the main focus of theorists of error statistics of experimentation such as Mayo—constitutes an important type of uncertainty, there are many more dimensions of uncertainty that are relevant in scientific practice. Of all typologies of uncertainty presented in the literature, the typology of uncertainty presented by Janssen et al. (2003) comes closest to my typology.[*] While some authors use the concept of uncertainty in a more narrowly defined sense, and distinguish it from the concept of 'risk'[†] (statistical uncertainty in my typology: we know the odds of events) and 'indeterminacy' (ontic uncertainty in my typology: variability), I have subsumed these concepts under a wider concept of uncertainty.[‡] The dimensions of methodological unreliability and value diversity are also included in the typology. However, the notion is not so broad that it encompasses 'pure ignorance' (we don't know what we don't know).[§]

I claim that this typology is applicable to all instances of scientific simulation. In Section II of this study, I illustrate how this typology can be used to characterise the uncertainties involved in simulating climate.

---

[*] The main difference between my typology and that of Janssen et al. (2003) is that I have added a separate dimension for recognised ignorance (see page 50, first note).

[†] In 1921, the economist Frank Knight introduced the still-influential distinction between *risk* (for which one can calculate the odds) and *uncertainty* (for which one cannot) (see Knight 2002).

[‡] Wynne (1992) distinguishes between 'risk', 'uncertainty', 'ignorance', and 'indeterminacy'.

[§] Wynne (1992: 114) states that 'a far more difficult problem [than uncertainty] is ignorance, which by definition escapes recognition'. Contrary to Wynne, I distinguish between recognised and pure ignorance and include recognised ignorance in my wider definition of uncertainty. I agree with Wynne, however, that recognised ignorance is the most difficult type of uncertainty.

4

Assessment of Simulation Uncertainty for Policy Advice

4.1 Introduction

The results of Chapters 2 and 3, which apply to the use of simulation in science, are also relevant for the provision of policy advice based on scientific simulations. Although most scientific simulations of nature are not used in policymaking, and policy advice is often given without using simulation, there are many important examples of policymaking that do rely on simulation outcomes. One of the prime examples is that of climate change, discussed in Section II of this study. Other examples are biodiversity loss; pollution of soil, water, and air; the design and management of nuclear weapons; the changes and spread of animal viruses (e.g., 'bird flu'); automobile passenger safety; and so on.

Society as a whole and policymakers in particular are often confronted with policy problems that involve significant uncertainties that are typically not diminished by the use of simulations. In fact, some of the questions asked of science in such cases 'cannot be answered by science' (Weinberg 1972); that is, even though the questions can be scientifically formulated, the uncertainties are too large to answer those questions unequivocally. Typically, many different answers can be produced by applying different simulation models to the policy issue. It thus becomes crucially important to know what can be expected from scientific advice and how the plurality of plausible models should be handled.

In this chapter, I study the process of 'assessment' of simulation uncertainty to provide information to policymakers and other societal actors involved in policy problems. I use the following definition of *assessment*:

> 'Assessment' is the analysis and review of information derived from research in order to help someone in a position of responsibility to evaluate possible actions, or to think about a problem. It does not usually entail doing new research. (van der Sluijs et al. 1998: 291)

I argue that by thoroughly assessing computer-simulation uncertainty and including reflection on the uncertainties in policy advice, simulation can play a meaningful role in policymaking. A similar argument was developed by Silvio Funtowicz and Jerry Ravetz (e.g., 1990, 1991, 1993). The aim of their work is to improve the decision-making process by introducing into the policy-advisory process appropriate information about the uncertainty and quality of the underlying science ('science providing advice to policy' can be called 'science-for-policy' in short*). In the Prologue to their book, this aim is set in the following context:

> There is a long tradition in public affairs which assumes that solutions to policy issues should, and can, be determined by 'the facts' expressed in quantitative form. But such quantitative information, either as particular inputs to decision-making or as general purpose statistics, is itself becoming increasingly problematic and afflicted by severe uncertainty. Previously it was assumed that Science provided 'hard facts' in numerical form, in contrast to the 'soft', interest-driven, value-laden determinants of politics. Now, policy makers increasingly need to make 'hard' decisions, choosing between conflicting options, using scientific information that is irremediably 'soft'. (Funtowicz and Ravetz 1990: 1)

In Funtowicz and Ravetz's analysis, the 'softness' of the scientific information about many pressing policy problems is a consequence of the fact that we cannot draw on knowledge gained from experiments but instead must use uncertain knowledge from simulation:

> Science cannot always provide well-founded theories based on experiments for explanation and prediction, but can frequently only achieve at best mathematical models and computer simulations, which are essentially untestable. Based on such uncertain inputs, decisions must be made, under somewhat urgent conditions. (Funtowicz and Ravetz 1991: 139; cf. Funtowicz and Ravetz 1990: 7)

In Chapter 3 of this study, it was concluded that it cannot generally be stated that simulation is less certain than experimentation. Hence, Funtowicz and Ravetz's suggestion that experiments always lead to certain knowledge ('well-founded theories') is wrong. Still, for the particular cases discussed by Funtowicz and Ravetz (e.g., very complex environmental issues) experimentation is impossible, and the question of the uncertainty of simulation is pressing indeed. Since policymakers are usually not themselves able to judge the uncertainty of scientific simulation-model

---

* The phrase *science for policy* was also used in the title of Funtowicz and Ravetz's work (1990): *Uncertainty and Quality in Science for Policy*.

outcomes, scientific policy advisers must carefully weigh how to present their conclusions.*

Simulation models of ecological systems, for example, although they may give an impression of how such systems behave and as such can suggest reasons for taking policy measures, cannot reliably predict the future states of these open and unpredictable systems. For policymakers and others involved in the policymaking process to be able to evaluate the simulation results, it is thus important that modelling assumptions are made transparent. In this chapter, it is argued that for reasons of acceptability to decision makers and of representativeness to different perspectives, it would be best if in concrete problem contexts, the relevant policy communities were involved in the framing of the models (what questions to address, where to put the system boundaries, etc.), in the choice of the models, and in the evaluation of the models. Simulation can then assist in the organisation of knowledge, stimulate mutual learning processes, and contribute to the integration of different perspectives on the problem (Haag and Kaupenjohann 2001: 57). Currently, this situation must still be regarded as an ideal, however. Simulation uncertainties do not often get the airing in policy advice that they deserve. One explanation is that scientific advisers are usually not asked by policymakers, politicians, and other actors to dwell on the uncertainties and treat them explicitly. For this reason, van Asselt and Petersen (2003) urged policymakers to be less 'afraid of uncertainty' and to stimulate policy advisers to communicate with them about uncertainties.

This chapter deals with the questions of why and how simulation uncertainty should be assessed and communicated in science advice to policy. Of course, not all areas of policymaking that involve advice derived from scientific simulation face the same level of uncertainty. After introducing the 'sound science' debate and the concept of 'postnormal science', a distinction between four types of policy problems will therefore be made. Simulation plays different roles in these different types of policy problems, and different demands are made on uncertainty assessment and communication. Subsequently, a general methodology for assessing and communicating simulation uncertainties in science-for-policy are outlined.

---

* Obviously, providing insight into the uncertainties involved in policy advice is necessary more generally, not only in the case of scientific computer simulation, but also in the case of experiments and observations. While the main emphasis in this chapter is on simulation-model uncertainty, the general discussion of the science–policy interface and the assessment of uncertainty in science-for-policy is not only valid for scientific simulation.

4.2 The Sound Science Debate

One area for which it is crucial that policy advisers carefully weigh their conclusions with respect to uncertainties is that of environmental policymaking. This can be demonstrated by looking at the American sound science debate, which has some parallels to the Dutch episode on the purported overreliance on simulation reported in Chapter 1.* In 1995, U.S. politicians—to be precise, the Republican majority of the House of Representatives—launched a stunning attack on the integrity of environmental scientists. In hearings before the Committee on Science, the politicians had given 'sceptics' on environmental issues, such as the ozone hole and human-induced global warming, a prominent role to deliver testimonies to the effect that the consensus statements by the majority of scientists on these issues were not based on sound science. The sceptics said that the incriminated environmental scientists had only been able to deliver model results that merely provided illusions about environmental problems. No real hard evidence had been given in the form of observations, according to the sceptics' judgements. Confirmed by these sceptics in their already-formed resolution to block the process of environmental policymaking, the House majority was able to pass legislation that demanded a basis in sound science (understood as based on observations alone) for any new environmental legislation.

Although the legislation did not make it through the Senate, the ranking minority member of the Committee on Science, the late Representative George E. Brown Jr., was deeply worried about this attack on the reliability of environmental science:

> This inordinate reliance on a single source of scientific understanding [i.e., observational data] is part of a broader view that the 'sound science' needed before regulation can be justified is science which somehow proves a proposition to be 'true'. This is a totally unrealistic view both of science's present capabilities and of the relationship between data and theory in the scientific method. Not coincidentally ... this approach to science can lead to near paralysis in policy making. (Brown 1996: 13)

Here Brown's worries concerned not only the political views of the Republicans (which he considered legitimate as views that could be aired in the political arena) but also even more their 'misunderstanding', as Brown called it, of science. Brown said he wanted to defend science against the kind of politicisation practised by the Republicans. According to Brown, politicians should trust the consensus that is presented by the majority of the

---

* The difference being that the U.S. politicians participated more fully in the debate and nearly took a decision to ban the use of computer simulation in environmental policymaking, while in the Netherlands the vast majority of politicians continued to trust the environmental scientists and their use of simulation.

scientists. He thought this trust could be granted to science since it arrives at this consensus through an 'objective' process of peer review mechanisms:

> Skepticism is an inherent part of the scientific perspective; scientific knowledge grows only through a process of continual questioning. The central accepted core of our scientific knowledge is the cumulative result of centuries of resolving scientific questions through observation, testing, and open and rigorous scientific peer review. The consensus that has emerged from this process deserves respect as our best effort to understand and explain the physical world. Like Winston Churchill's democracy [which is the least-bad form of government], peer-review is not a perfect process, but it is the best that we have. (Brown 1996: 17)

The Republicans who were attacked by Brown showed 'a systematic aversion to the use of theory, models, and other sources of scientific knowledge to provide a full understanding of observed data' (p. 13), at odds with their role in scientific practice, as confirmed through the mechanism of peer review, according to Brown (1996). From the analysis presented in this study, we must conclude that Brown relied too much in his statement on the possibility of arriving at consensus on the simulation of complex environmental systems. However, he was right about the fact that modelling is pervasive in science.

Some philosophers of science have taken a clear stance in the sound science debate and supported the Democrats' criticism of the Republicans. One line of argument is to claim that there are no philosophically relevant differences between simulation and experimentation, and hence that the presentation by the Republicans of experiment and simulation as opposites is misguided. Here is, for example, the central claim of a publication by Stephen Norton and Frederick Suppe:

> Simulation modeling is just another form of experimentation, and simulation results are nothing other than models of data. Thus, the epistemological challenges facing simulation models are, at bottom, identical with those of any other experimental situation involving instrumentation. (Norton and Suppe 2001: 92)

I disagree with this claim, however. Norton and Suppe overemphasise the similarities between simulation and experimental practice, which can partly be attributed to the case study they analysed (involving a simulation model as part of a measurement apparatus). Yet the sources of uncertainty in the two practices are not the same, as was demonstrated in Chapter 3. If philosophers want to provide a useful contribution to the sound science debate, they will need to use other arguments. My strategy is to show that although the uncertainties associated with simulating complex environmental problems are too large to answer many policy questions unequivocally, by conscientiously assessing and communicating these uncertainties (and comparing them with the large uncertainties associated with observational data), a sensible use of simulation results in policymaking is possible.

4.3 The Challenge of Postnormal Science

Different views exist on the role of science in policy. Yaron Ezrahi (1980) identifies two main views: the 'utopian rationalist' and the 'pragmatic rationalist' view. The utopian rationalist, or 'technocratic', ideal of science advice entails that the policymaking process should assimilate scientific information to a maximum extent. This ideal reflects the notion of science speaking 'value-free' truth to political power that gained institutional currency in the 19th century. The notion of value-free science itself was based on the expectation that the impartiality and objectivity of scientists could help to overcome political conflict (Proctor 1991). In the 20th century, however, it became clear both that science cannot be value free and that politics deals increasingly often with issues that are clouded with uncertainty, including value diversity. The presence of conflict among scientists, both epistemic and social, makes it hard to provide politicians with neutral advice. As was argued in Chapter 3, there is often considerable room for scientists and policy analysts to make choices in the assumptions of the analysis; for example, when simulations are used, simulationists make choices in the conceptual and mathematical model, the model inputs, the technical model implementation, and the processing of the output data and their interpretation.

Apart from the utopian rationalist model, Ezrahi identifies a pragmatic rationalist, or 'democratic', ideal of science advice that accepts, within limits, the inevitability of political ingredients in science advice. Pragmatic rationalism considers technocratic policies to be fallible. Science can then contribute to political debate by representing different legitimate perspectives on policy problems. According to Ezrahi, the two models of science advising are not mutually exclusive, but each model is most fruitfully applied in different types of policy problems.* From the perspective of those favouring technocratic politics, the requirement that policy problems should become democratised may seem to be a hindrance for policymaking. But from the perspective of those who favour participatory politics and a more reflective way of defining policy problems, the political condition in which we find ourselves when science cannot be 'speaking truth to power' (Price 1965) may be considered an opportunity for the democratisation of both the scientific advisory process and politics in general. This democratisation could start from a renewed awareness that it is inevitable that political decisions will be made under conditions of uncertainty:

---

* Ezrahi's different types of policy problems are introduced in the next section.

> To be political is to *have* to choose—and, what is worse, to have to choose
> under the worst possible circumstances, when the grounds of choice
> are not given a priori or by fiat or by pure knowledge (*episteme*). To be
> political is thus to be free with a vengeance—to be free in the unwel-
> come sense of being without guiding standards or determining norms,
> yet under an ineluctable pressure to act, and to act with deliberation and
> responsibility as well. (Barber 1984: 121)

Scientists—or philosophers, for that matter—cannot deduce the norms
that should govern political decision making from empirical research or
pure thought. The key concepts of democratic politics are 'deliberation' and
'responsibility'. Scientists and philosophers can, of course, take part in these
deliberations, but they have a responsibility to consider the perspectives of
other citizens.

Sheila Jasanoff, in her study of the role of several American advisory committees on risks (Jasanoff 1990), rejects both the technocratic model and its
typical democratic critique. The technocrats argue for deference to the scientific community when technical matters have to be decided (or alternatively,
they claim that technical material used in policymaking must undergo rigorous peer review by the scientific community). The main problem with their
conception of scientific advice is that technical issues typically cannot be
depoliticised without causing controversy since many of the issues are value
laden. The democrats, on the other hand, argue for a better incorporation of
different societal values in science-based decision making. They demand that
membership of scientific advisory committees should be opened to include
different—not necessarily technical—viewpoints. Jasanoff concludes from
her study of scientific advisory committees connected to U.S. administrative
agencies that 'it is crucial for claims certified by agency advisers to be per-
suasively labelled "science"' (Jasanoff 1990: 244). The ideal scientific adviser
is an esteemed scientist, who 'not only transcends disciplinary boundaries
and synthesizes knowledge from several fields but also understands the
limits of regulatory science and the policy issues confronting the agency'
(Jasanoff 1990: 243). From a democratic point of view, the precondition of the
legitimacy of scientific policy advice then is that the advisers, who necessarily exercise political judgement when they give scientific advice, must act as
citizens who are free to make their own choices (cf. Barber 1984: 127). The
advisers must thus be responsive to the fact that different societal and political actors may have different perspectives on the underlying uncertainties,
and it is important that they integrate these perspectives within the advisory
process. As a corollary, interest groups should not be directly represented in
scientific advisory committees.* According to Jasanoff, instead of speaking
truth to power, scientific advisers may still hope to deliver what she calls
'serviceable truth', which she defines as advice that is, on the one hand, sci-

---

* Jasanoff's (1990) position thus comes closer to the technocratic than to the democratic ideal.

entifically acceptable and able to support reasoned decision making and, on the other hand, reflects the scientific uncertainties.

How scientists and policymakers interact at the interface between science and policy has been studied empirically in terms of the 'boundary work', through which the boundary between science and policy is maintained (e.g., Gieryn 1999; Jasanoff 1990). The main conclusion of these studies is that it is impossible to find stable criteria that absolutely distinguish science from nonscience (e.g., politics). Many social scientists who have studied the relationship between the practices of science and decision making have indeed concluded that these two categories of activities cannot be neatly separated (e.g., Jasanoff and Wynne 1998).

An example may serve to illustrate this point. The ongoing scientific assessment process of the climate change issue that is conducted by the Intergovernmental Panel on Climate Change (IPCC) receives questions from and feeds back into the United Nations Framework Convention on Climate Change (UNFCCC). Due to widely publicised warnings from scientists in the 1980s, the public in Western democracies became interested in the risks involved in an enhanced greenhouse effect induced by anthropogenic emissions of CO_2, leading to human-induced global warming and its associated effects, such as sea-level rise. The attribution of climate change to human influences and the projections of climate change into the future have made heavy use of climate simulations. Since the societal changes implied by the different solutions proposed for solving the global warming problem are quite drastic, one of the first steps politicians took to address the problem was to ask scientists to assess the state of climate science regularly, as well as the possibilities for societal adaptation to climate change and mitigation of the problem by reducing anthropogenic greenhouse gas (mostly CO_2) emissions. This led in 1988 to the establishment of the IPCC.[*] The advisory process involving the IPCC is regarded by many social scientists as a 'coproduction' of, on the one hand, our knowledge about the climate system and, on the other hand, the international political order. This is reflected by the institutional ties between the IPCC and the UNFCCC and by the kind of knowledge that is produced by the IPCC:

> The IPCC's efforts to provide usable knowledge resonated with the belief of sponsoring policy organizations that climate change is a manageable problem within the framework of existing institutions and cultures. (Jasanoff and Wynne 1998: 37)

---

[*] The IPCC consists of three working groups. The current distribution of subjects addressed by each working group is as follows: Working Group I deals with the (natural) scientific basis of climate change; Working Group II addresses issues of impacts, adaptation, and vulnerability; and Working Group III assesses mitigation options. The analysis presented in Section II of this study focuses on Working Group I.

The processes that lead to usable knowledge being delivered by the IPCC and involve boundary work are closely scrutinised in Chapter 7.

The mutual reinforcement of scientific assessment and political decision making seems to be a common feature of environmental assessment. A second example of this phenomenon comes from an analysis by Peter Haas, who has studied the development of the Mediterranean Action Plan, or 'Med Plan'. The Med Plan was agreed on in 1975 and was implemented in the succeeding decades. It is a regional environmental cooperation for dealing with the issue of marine pollution in the Mediterranean. The information underlying the Med Plan suffers from uncertainties in ecotoxicological simulations, among other uncertainties. The main scientists and policymakers involved in the Med Plan, however, 'shared an abiding belief in ecological principles and were committed to preserving the physical environment, which they thought was threatened by pollution' (Haas 1990: 74–75). These ecological principles were partly derived from theoretical ecological computer simulations that are used to study the behaviour of complex ecological systems. The significant uncertainties associated with these simulations are hardly aired in the convention and protocols of the Med Plan.[*] They were dealt with by the actors involved at an unreflective level. Haas shows that an 'ecological epistemic community' of professionals had been involved in both the scientific assessments and the formation of the plan.[†] An epistemic community consists of professionals who believe in the same cause-and-effect relationships and share common values (Haas 1990: 55). In the 1970s, system-dynamical metaphors about ecology had gained currency in the political domain (cf. Kwa 1987). The shared beliefs in ecological principles were thus spread outside science, facilitating the mutual reinforcement of scientific assessments and policymaking in the example of the Med Plan. This is illustrated by the fact that during the early negotiations at the beginning of the 1970s, the membership of the epistemic community included, aside from the leadership of the United Nations Environment Programme (UNEP):

> members of the Greek government, French modelers and systems scientists, UNESCO [United Nations Educational, Scientific and Cultural Organization] bureaucrats, FAO [Food and Agriculture Organization] lawyers, and individuals in the Israeli, Spanish, and Egyptian governments. (Haas 1990: 75)

The professionals who shared their beliefs in ecological principles were thus located within and outside science. The authority of the scientists within this epistemic community was not questioned by actors outside the epistemic community, even though the uncertainties were large. According

---

[*] See http://www.unepmap.org.
[†] One can thus say that the scientific assessments and the Med Plan were coproduced.

to Haas, epistemic communities are particularly effective in influencing policymaking in issue areas in which uncertainty is high:

> As these types of issues with a high degree of uncertainty gain in salience for leaders—as is indeed the case—there is a greater range of influence for epistemic communities that possess authoritative claims to understanding the problems; analyzing them is also more useful. (Haas 1990: 246)

According to Haas, there were no industry challenges to the decisions made by the experts, mainly because at the time the Med Plan was agreed the countries of the European Economic Community (EEC) already had an EEC directive in place. Furthermore, the public was not very interested in marine issues (Haas 1990: 163). In sum, in the case of the Med Plan there was no perceived need to address simulation uncertainties explicitly in the policy process.

In international environmental policymaking, the explicit acknowledgement of uncertainty in the science that is used to underpin policy has become more frequent since the 1990s. The rise to prominence of the 'precautionary principle' marks a reflective transition of governments' attitudes toward scientific uncertainty. The World Commission on the Ethics of Scientific Knowledge and Technology (COMEST) of UNESCO offers the following 'working definition' of this principle:

> When human activities may lead to morally unacceptable harm that is scientifically plausible but uncertain, actions shall be taken to avoid or diminish that harm. (UNESCO 2005: 14)

The principle urges politicians to take measures even when uncertainty about a problem still exists, provided that scientific analysis has taken place and uncertainties have been thoroughly assessed.[*] From the end of the 1980s, the significant scientific uncertainties surrounding large-scale and high-impact environmental problems, such as biodiversity loss and climate change, started to become explicitly referred to in policy documents, in combination with references to the precautionary principle.

The special challenges facing experts under conditions of high societal stakes and high scientific uncertainty, such as those that have become evident in the area of climate change, were also identified in the area of risk

---

[*] Contrary to conventional wisdom that sees the European Union endorsing the precautionary principle and proactively regulating risks and the United States opposing it and waiting for evidence of harm before regulating, transatlantic comparisons have demonstrated that differences in relative precaution depend more on the context of the particular risk than on broad differences in national regulatory regimes (Wiener and Rogers 2002).

assessment in the mid-1980s.* Recognising that the interactions between science and policymaking on risks are often unproductive when the decision stakes and system uncertainty are very high (in the case of nuclear energy, for example), Funtowicz and Ravetz propose distinguishing a new type of risk assessment called 'total-environmental assessment' (Funtowicz and Ravetz 1985: 228). This is a form of risk assessment in which the 'total environment'—that is, the complete context—of a risk issue is taken into account as much as possible. This kind of risk assessment is appropriate for cases with high decision stakes and system uncertainty.† Funtowicz and Ravetz (1985: 228–229) describe the 'methodology' of total-environmental assessment as follows:

> [It] is permeated by qualitative judgments and value commitments. Its result is a contribution to an essentially political debate on larger issues, though no less rational in its own way for that. The inquiry, even into technical questions, takes the form largely of a dialogue, which may be in an advocacy or even in an adversary mode.

In very polarised settings, the least one can hope for, according to Funtowicz and Ravetz (1985: 229), is a 'consensus over salient areas of debate'.

According to Funtowicz and Ravetz, structural changes in the direction of enhanced participation are needed to democratise scientific advisory proceedings. For this reason, they have generalised their original normative view on risk assessment into a sweeping normative statement on the future of science-for-policy:

> Now global environmental issues present new tasks for science; instead of discovery and application of facts, the new fundamental achievements for science must be in meeting these challenges. ... In this essay, we make the first articulation of a new scientific method, which does not pretend to be either value-free or ethically neutral. The product of such a method, applied to this new enterprise, is what we call 'post-normal science'. (Funtowicz and Ravetz 1991: 138)

---

* *Risk* is used here as 'a concept to give meaning to things, forces, or circumstances that pose danger to people or to what they value. Descriptions of risk are typically stated in terms of the likelihood of harm or loss from a hazard and usually include: an identification of what is "at risk" and may be harmed or lost (e.g., health of human beings or an ecosystem, personal property, quality of life, ability to carry on an economic activity); the hazard that may occasion this loss; and a judgment about the likelihood that harm will occur' (U.S. National Research Council 1996b: 215–216). This definition of risk leaves room for pluralism: People can have different perspectives on risks. For instance, whether someone considers something to be at risk depends on their valuation of what could be at risk.

† These two variables are not totally independent in the sense that the recognition of system uncertainty is typically enhanced if the decision stakes are high (see Jasanoff and Wynne 1998: 12).

While Funtowicz and Ravetz first wrote about 'risk assessment', they thus subsequently applied their analysis to 'problem-solving strategies' more generally. The problem-solving strategy that they call postnormal science (or 'second-order science') corresponds to the total-environmental method of risk assessment applied to global environmental issues, among other problems (Funtowicz and Ravetz 1991: 137, 144–145).*

Which institutions could facilitate the application of the strategy of post-normal science to complex policy problems? In the literature, the IPCC has been identified as comprising elements of postnormal science (Bray and von Storch 1999; Saloranta 2001). The IPCC has also been considered as a 'boundary organisation' (Guston 2001; Miller 2001). David Guston provides, as a definition of *boundary organisations*, that they meet the following three criteria:

> [F]irst, they provide the opportunity and sometimes the incentives for the creation and use of boundary objects and standardized packages; second, they involve the participation of actors from both sides of the boundary as well as professionals who serve a mediating role; third, they exist at the frontier of the two relatively different social worlds of politics and science, but they have distinct lines of accountability to each. (Guston 2001: 400–401).

Boundary objects are conceptual or material objects sitting between two different social worlds, such as science and policy, and they can be used by individuals within each for specific purposes without losing their own identity (Star and Griesemer 1989). An example is 'climate sensitivity' (the sensitivity of climate to perturbation by greenhouse gases—defined as the temperature change resulting from a doubling of the CO_2 concentration—that can be determined using climate data and climate simulation). Climate modellers use the concept of 'climate sensitivity' as a benchmark for comparing their models. Climate modellers who use simple models often use the climate sensitivity simulated by more complex models as a model parameter. And for policymakers and advisers, climate sensitivity provides a 'window' into the world of climate modelling (van der Sluijs et al. 1998: 310). Surprisingly, given the large uncertainties associated with determining climate sensitivity, the uncertainty range has remained constant at 1.5–4.5°C since the first assessment of climate sensitivity by the U.S. National Academy of Sciences in 1979. As van der Sluijs et al. (1998) show, many different interpretations have been given to this range, both statistical and nonstatistical (scenario: 'what-if') interpretations, which prompted them to call this boundary object an 'anchoring device' and study the social causes of retaining the consensus range of 1.5–4.5°C. 'Standardised packages' are more broadly defined than boundary objects. They 'consist

---

* The other two types of problem-solving strategies are applied science (low systems uncertainty or low decision stakes) and professional consultancy (medium-level systems uncertainty or medium-level decision stakes) (e.g., Funtowicz and Ravetz 1991, 1993).

of scientific theories and a standardised set of technologies or procedures and as a concept handle both collective work across divergent social worlds and 'fact stabilisation' (Fujimura 1992). To take again the example of climate change, standardised packages can be found in the conceptualisation of climate change and the establishment of a thriving line of climate research in coordination with climate policymaking.

An alternative definition of boundary organisations is provided by Clark Miller, who pleads that, especially in the study of the boundary between science and politics at the international level, we should not focus on structure but on process and dynamics. According to Miller, boundary organisations are organisations that take part in 'hybrid management', with *hybrids* being

> social constructs that contain both scientific and political elements, often sufficiently intertwined to render separation a practical impossibility. They can include conceptual or material artifacts (e.g., the climate system or a nuclear power plant), techniques or practices (e.g., methods for attributing greenhouse gas emissions to particular countries), or organizations (e.g., the SBSTA [scientific and technological body of the climate convention] or the Intergovernmental Panel on Climate Change). (Miller 2001: 480)

Hybrid management activities are not necessarily limited to work carried out in boundary organisations.

In this chapter, I show that the Netherlands Environmental Assessment Agency can be considered an example of an organisation that is both structured as a boundary organisation (Guston) and features processes of hybrid management (Miller). The latter becomes evident from the adoption of specific procedures to take up the challenge of postnormal science (Funtowicz and Ravetz), that is, of providing responsible scientific policy advice under conditions of high stakes and high uncertainty, as described in this chapter. In Section II of this book, the IPCC is also shown to have the structure of a boundary organisation and the features of hybrid management. My general claim is that boundary organisations between science and policy can be effective institutions to take up the challenge of postnormal science (see also Petersen et al. 2011), but that it depends on the specific procedures adopted whether the assessment and communication of simulation uncertainty are done adequately and responsibly.

4.4 The Role of Simulation Uncertainty in Policy Advice

From studies in political science, one can conclude that not all instances of science-for-policy are as dramatic as the postnormal science cases of the complex environmental problems addressed by Funtowicz and Ravetz. The way simulation uncertainties should be dealt with differs according to the policy-problem types; therefore, it is important for both scientific advisers and policymakers to reflect on the type of policy problem that they are facing. The kind of boundary work and the level of interaction between science and policy differs according to (1) the level of scientific uncertainty and (2) the level of societal and political debate. The political scientist Yaron Ezrahi (1980) distinguishes between four types of policy problems, varying in the level of agreement on the political objectives and on the scientific knowledge relevant to the problem. In a particular policy domain, there could be a situation of

1. agreement on political objectives with scientific consensus;
2. agreement on political objectives without scientific consensus;
3. scientific consensus and disagreement about political objectives; and
4. disagreement about political objectives coupled with scientific dissensus.

Figure 4.1 shows a diagram featuring Ezrahi's four types of policy problems, with some additional characterisations proposed by Hisschemöller and Hoppe (1996) and Hisschemöller et al. (2001). For each type of policy problem, one example is given in the diagram of a policy area in which simulations of nature play a significant role.

When agreement is assumed on the societal norms and values and on the required kind of knowledge in addressing the policy problem, the problem can be called 'structured'. Scientific advisers are then taken to be able to provide unequivocal information to policymakers, which can unproblematically be used in the policy process (the utopian rationalist model). An example of a structured policy problem that involves scientific simulations is automobile safety regulation. In the case of automobile safety, simulations are used in the execution of safety policy. For example, until the mid-1980s, regulators only allowed automobile manufacturers to meet legal safety requirements by performing laboratory crashes on instrumented production cars, at a cost of hundreds of thousands of euros per test. Since the advent of reliable computer simulations of car crashes, most of world's automobile manufacturers have supplemented their laboratory crash tests with supercomputer simulations (Kaufmann and Smarr 1993: 157–159).

At the other extreme, when there is acknowledgement of a clash of values related to the problem and disagreement on what scientific knowledge should be used to solve the problem, the problem is called 'unstructured', and scientists may only serve as 'problem recognisers' (the pragmatic rationalist model). In this situation, scientific advisers must base their authority on their ability to assess and communicate uncertainty and analyse the different

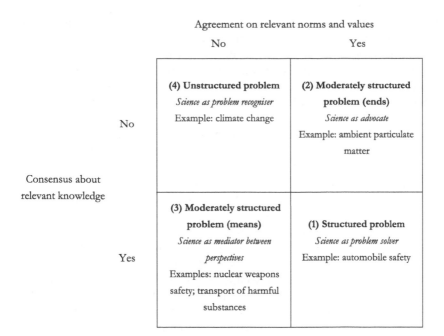

Agreement on relevant norms and values

	No	Yes
No	**(4) Unstructured problem** *Science as problem recogniser* Example: climate change	**(2) Moderately structured problem (ends)** *Science as advocate* Example: ambient particulate matter
Yes	**(3) Moderately structured problem (means)** *Science as mediator between perspectives* Examples: nuclear weapons safety; transport of harmful substances	**(1) Structured problem** *Science as problem solver* Example: automobile safety

Consensus about relevant knowledge

FIGURE 4.1

Four types of policy problems with associated roles. (Diagram with roles from Hisschemöller, M., Hoppe, R., Groenewegen, P., and Midden, C.J.H. (2001), Knowledge use and political choice in Dutch environmental policy: a problem-structuring perspective on real life experiments in extended peer review, In M. Hisschemöller, R. Hoppe, W.N. Dunn, and J.R. Ravetz (eds.), *Knowledge, Power, and Participation in Environmental Policy Analysis*, New Brunswick, NJ: Transaction, pp. 437–470: 464. The numbers refer to Ezrahi's list, see this book on page 78. The examples have been added by the author.)

values and perspectives on the problem.[*] Again, this is pictured here as an ideal; it is not to say that scientific advisers can easily do this or that they need to do it alone without involving policymakers, politicians, stakeholders, and citizens. The issue of climate change, discussed in Section II of this study, constitutes a prime example of this type of policy problem in which policymakers extensively rely on computer simulations of the climate system.

The typology of policy problems shown in Figure 4.1 also allows for 'in-between' problems, which are called 'moderately structured' problems (either knowledge or values are considered to be structured for such problems).

---

[*] Roger Pielke (2007) distinguishes between four idealised roles scientists may choose from when advising policy: the roles of Pure Scientist (who seeks only the truth without considering the practical implications of his or her research), Science Arbiter (who seeks to focus on issues that can unequivocally be resolved by science), Issue Advocate (who seeks to advance particular interests using his or her expert status), or Honest Broker of Policy Alternatives. The last role is applicable in the case of policy problems with large scientific uncertainties and high societal stakes (of the unstructured type).

Policies related to, for instance, ambient particulate matter ('fine dust'), which involve simulations of the spread and health risks resulting from the emission of polluting substances, constitute an example of policies assuming consent on the values and norms ('ends'). On the one hand, the end goal of the protection of human health is relatively undisputed, but on the other hand, policymakers have to deal with large uncertainties about emissions, human exposure, and the attribution of causal effects to individual species of particles or other pollutants (Maas 2007; van der Sluijs et al. 2008; Knol et al. 2009).

An example of a moderately structured policy problem for which there is consent on the possible solutions (means) but widespread disagreement on the political goals is the U.S. policy for maintaining the safety of nuclear weapons in the stockpile. Although the United States has ceased nuclear testing, nuclear deterrence remains an integral part of U.S. defence policy, and the United States plans to maintain an arsenal of several thousand nuclear weapons for the foreseeable future. America's weapons laboratories—Los Alamos, Lawrence Livermore, and Sandia—must now certify the reliability of old weapons and train new weapons scientists without nuclear testing. Toward that end, the Clinton administration initiated a multifaceted programme of 'science-based stockpile stewardship'. Despite initial concerns about the reliability of using computer simulations to determine the safety of the bombs in storage (see Petersen 1999a), the Advanced Simulation and Computing program, established in 1995, stimulated the development of ever-faster supercomputers—in November 2010, the U.S. nuclear weapons laboratories owned four of the world's ten fastest supercomputers, from both Cray and IBM[*]—and provided partially validated simulation models from 2004 onwards to support the annual certification of the stockpile and to assess manufacturing options. Several warheads have now actually been recertified on that basis.[†] These simulation models are judged reliable enough for the certification of existing weapons. By proposing and subsequently developing a 'virtual' testing capacity, scientists took on the role of 'mediator' between policy actors who advocated the importance of maintaining nuclear arsenals and policy actors who advocated a comprehensive nuclear test ban (and who were not necessarily in favour of maintaining nuclear weapons).

To give a further example of the mediating role for scientific advisers, in 2003 the Netherlands Environmental Assessment Agency provided mediating advice to the Dutch government in the area of risk management. The government had encountered several risks that it had found difficult to decide on in a structured policy-problem mode. Examples were risks arising from (1) storage, use, and transport of harmful substances, such as liquid natural gas; (2) airplane accidents; and (3) exposure to *Legionella*. Results from simulations are often used in the respective risk assessments, which therefore have

---

[*] See http://www.top500.org/lists/2010/11 (accessed 5 June 2011).
[†] See http://www.sandia.gov/NNSA/ASC/programs/progs.html (accessed 5 June 2011).

some simulation uncertainty associated with them.* However, in these risk assessments the scientific uncertainties are not considered very policy relevant by the risk assessors. It is rather the societal values that are considered to be at stake. For instance, policymakers were reminded in the report that when the cost of guaranteeing every Dutch resident a particular protection level (determined by policy choices) is very high, the political decision can be taken either to search for less-expensive forms of risk reduction or to accept a greater risk for specific risks (National Instititute for Public Health and the Environment/Netherlands Environmental Assessment Agency [RIVM/ MNP] 2003: 6). The rationale a government can give for accepting higher risks in a particular case is that the qualitative, sociopsychological characteristics of risk are intrinsic components of the 'risk' concept, besides the likelihood of harm or loss.† Thus scientific advisers who take on a mediating role can, through their risk assessments, contribute to mediation between the different social values pertaining to the risks. In principle, the actual mediation in decision making is not carried out by the scientists, but by the legitimate decision makers. It is important for risk assessors to be aware of the fact that feelings of citizens, and politicians, play a legitimate role in public decision making on risk management (Slovic et al. 2004).‡ In risk assessment, there are thus roles for both scientists and societal stakeholders:

> [C]riteria for evaluating risks should be developed from the social discourse about concerns, while the 'objective' measurement should be performed by the most professional experts at hand. (Klinke and Renn 2002: 1077)

If the 'facts' are also uncertain, we face an unstructured problem type.

How to evaluate and manage risks should depend on the relative degrees of complexity, uncertainty, and ambiguity of the risks (Klinke and Renn

---

* These simulations pertain to many aspects of the problems. Examples are (1) explosions of liquid natural gas, (2) flight movements, and (3) water treatment. Many of these simulations concern not only simulation of nature but also simulation of human behaviour.

† See the definition of risk given on page 75, first note.

‡ There is a continuous interaction between reasoning grounded in feelings and reasoning grounded in universally applicable rules (cf. Brown's model of rationality introduced in Chapter 1). Scientific advisers must be aware of the fact that their judgements may be incomplete if they do not take the reasons of citizens for their evaluation of risks into account. In addition, it is important for advisers to realise that they are actors who may influence the feelings of citizens by the way the results of risk assessments are presented; in the case of risk communication, for instance, it has been observed that the feelings caused by the presentation of risks depend on whether the risks were communicated as percentages (e.g., 0.1%) or as relative frequencies (e.g., 1 in 1,000). The latter way of communicating risk leads to stronger feelings; the strongest feelings are caused by individual stories. To prevent the use of simulation models in public policymaking from neglecting 'softer' aspects of the policy problems, for which, for example, qualitative uncertainty assessment is needed, policy analysts, in addition to using models, must draw on their feelings, according to the psychologist Paul Slovic.

2002).* According to sociologists Andreas Klinke and Ortwin Renn, three strategies for dealing with risks are at our disposal: the classical risk-based strategy in which we are able to work with numbers that characterise the risk (Ezrahi problem type 1), the precaution-based strategy in which we prudently deal with uncertainty and vulnerability (Ezrahi problem types 2 and 4), and the (participatory) discourse-based strategy needed when values are in dispute (Ezrahi problem types 3 and 4). Obviously, discourse, both cognitive and reflective, also plays a role in combination with the first two strategies. Stakeholders must be involved in making decisions on how to classify a risk:

> Obviously, one needs a screening exercise to position the risk in accordance with our decision tree[†] and to characterize the degree of complexity, uncertainty and ambiguity associated with the risk under investigation. We would recommend a 'risk characterization panel' consisting of experts (natural and social scientists), some major representatives of stakeholders, and regulators who will perform this initial screening. (Klinke and Renn 2002: 1091)

This requires from regulators that they guard themselves against assuming too quickly that a particular problem belongs to the structured type (Ezrahi problem type 1) since other actors may strongly disagree with that categorisation.

The typology of policy problems presented in Figure 4.1 serves to identify ideal types of policy problems and their implications for appropriately assessing and communicating uncertainties. One must be aware that, in practice, in the course of dynamic policy processes, the policymakers' and others' views of problems may shift into another problem type. This means that the typology of policy problems shown in Figure 4.1 should not be regarded as implying a rigid categorisation of policy problems. Also at any given moment in time, the categorisation is not necessarily unequivocal. Although one of the problem types may dominate the practice of policy-making, different actors often hold different views on the categorisation of a problem. Thus, typically, all strategies associated with the different problem types can be observed. Due to the dynamics of policy problems, dominant strategies and roles may shift.

Consider the nuclear weapons safety policy as an example of a change of policy-problem type over time. Since the virtual testing capacity can also

---

* *Ambiguity* is the term used by Klinke and Renn (2002) to denote differences in how problems are framed by different actors. In my typology, ambiguity is related to the value-diversity dimension of uncertainty.
† Klinke and Renn (2002: 1083) present a 'decision tree' for classifying risks into one of the six classes that they distinguish. I do not discuss their classes here, but merely list the five main questions of their decision tree: (1) Is the risk potential known? (2) Thresholds on criteria exceeded? (3) Damage potential known? (4) Disaster potential high? and (5) Social mobilisation high?

be used for developing *new* nuclear weapons, the question arises whether new weapons will also be certified without real tests. Thus, scientific uncertainties around nuclear weapons policies increase, and the policy problem moves to the unstructured category. Indeed, the current U.S. wish to keep an underground nuclear test readiness* can be interpreted as a judgement on the limited reliability of simulating new nuclear weapons designs. Of course, the political goal of developing new nuclear weapons is even more controversial than the goal of maintaining existing stockpiles.

Collingridge and Reeve (1986) have given examples of policy problems in which the assessment by the policymakers and scientists of the policy type differed. Sometimes, the expectations of scientists about their influence on the policy process are set too high. From the technocratic viewpoint, science often seems to have only a negligible influence on policymaking. Collingridge and Reeve consider two modes for the relation between science and policy. First, a political consensus already exists, and supportive scientific evidence is selectively used to legitimate policy, applying strategies related to Ezrahi's type 2 problem and using scientists as advocates, while the scientists aspire to be problem solvers (type 1 problem). Second, political opponents keep arguing about technical details, continually deconstructing each other's scientific claims and not allowing external expertise to independently establish the plausibility of the different claims. The policy issue is then framed as an unstructured problem (type 4), while the advisers may again assume that they are dealing with a structured problem (type 1).

Now that we have identified the different roles simulation is able to play in policymaking, we can determine which types of uncertainty could be expected to play a significant role in the policy process. Advisers should give priority to assessing and communicating particularly those uncertainties. The following types of uncertainty are important in the settings of the different policy-problem types (Janssen et al. 2003: 11; van der Sluijs et al. 2003: 11–12):

1. If the problem is structured, statistical analyses and reporting of uncertainty ranges are typically suitable strategies;

2. If the problem is moderately structured (ends), weaknesses in the knowledge base and recognised ignorance become very relevant;

3. If the problem is moderately structured (means), the value-ladenness of assumptions and scenario uncertainty particularly need to be addressed;

---

* To retain the readiness to resume nuclear testing constitutes part of the scientific efforts within the defence programs of the National Nuclear Security Administration (see http://nnsa.energy.gov/aboutus/ourprograms/defenseprograms/futurescienceandtechnologyprograms/science, accessed 5 June 2011).

4. If the problem is unstructured, recognised ignorance, weaknesses in the knowledge base, and value-ladenness of assumptions all come to the fore.

Particularly in the last type of policy problem (type 4, the unstructured problem), it is difficult for scientists to assess and communicate uncertainties systematically; they are generally not trained to deliver advice under such circumstances. Funtowicz and Ravetz observe that when the stakes and uncertainties are high (and hence the uncertainties are amenable to politicisation), that 'there are at present no mechanisms towards a consensus on such politicized uncertainties' (Funtowicz and Ravetz 1990: 15). With respect to the possible solution that could follow from improved management of uncertainty within science-for-policy, they modestly conclude:

> Any genuine attempt to improve the quality of scientific information as it is used in the policy process must be undertaken with such political realities in mind. (Funtowicz and Ravetz 1990: 16)

In the following section, a methodology to assess and communicate the appropriate uncertainties is offered.

4.5 The *Guidance on Uncertainty Assessment and Communication*

Governmental and intergovernmental agencies that provide scientific advice to policymakers increasingly recognise that uncertainty needs to be dealt with in a transparent and effective manner. The Netherlands Environmental Assessment Agency (Milieu-en Natuurplanbureau [MNP] until 2008 and Planbureau voor de Leefomgeving [PBL] since 2008) is an example of an institution that interfaces science and policy by performing independent scientific assessments and policy evaluations, and that has developed a guidance for how to deal with uncertainty. The PBL Netherlands Environmental Assessment Agency is organised as a boundary organisation, which becomes evident from

1. the use of a standardised set of models that aim to integrate state-of-the-art scientific knowledge and to be applicable to the evaluation of policy proposals;
2. the participation of scientists and policymakers in the production of assessment reports, the position of the agency within the government bureaucracy (formally, the agency is a part of the Ministry of Infrastructure and the Environment), the presence of the PBL director at cabinet meetings; and

3. regular scientific reviews of the quality of PBL work and consultation with the government on the policy relevance of the annual work plan, thus providing different lines of accountability to science and policy.

The PBL is engaged in hybrid management; its assessments contain both scientific and political elements. One area in which this becomes visible is delineated by the procedures that the PBL has adopted for uncertainty assessment and communication.

The Dutch Environmental Protection Act (Wet milieubeheer) has since 1994 contained a clause that demands that the Netherlands Environmental Assessment Agency assess and communicate uncertainties in its environmental outlook reports.* In response to the media affair described in Chapter 1, the Netherlands Environmental Assessment Agency has implemented a comprehensive, multidisciplinary approach to uncertainty assessment and communication that applies to all types of assessments produced by the agency. This approach takes into account the societal context of knowledge production and constitutes a major conceptual and institutional innovation.

Since the scientific assessments produced by agencies such as PBL have to integrate information covering the entire spectrum from well-established scientific knowledge to educated guesses, preliminary models, and tentative assumptions, uncertainty cannot generally be remedied through additional research or comparative evaluations of evidence by expert panels searching for a consensus interpretation of the risks. The social studies of scientific advice mentioned in the previous sections show that for many complex problems, the processes within the scientific community, as well as between this community and the 'external' world (policymakers, stakeholders, and civil society), determine the acceptability of a scientific assessment as a shared basis for action. These processes concern, among other things, the framing of the problem, the choice of methods, the strategy for gathering the data, the review and interpretation of results, the distribution of roles in knowledge production and assessment, and the function of the results in the policy arena. Although assumptions underlying the design of these processes are still rarely discussed openly within or outside PBL, they are important for the knowledge produced by the agency becoming either 'contested' or 'robust'.†

The PBL acknowledges that it is not enough to analyse uncertainty as a 'technical' problem or merely seek consensus interpretations of inconclusive evidence. In addition, the production of knowledge and the assessment of uncertainty have to address deeper uncertainties that reside in

---

* See Wet milieubeheer article 4.2.1 (concerning PBL's environmental outlooks), effective since 1994, originally for Rijksinstituut voor Volksgezondheid en Milieu [RIVM], but since 1 January 2006 for MNP (succeeded by PBL in May 2008) as a separate organisation.

† In a recent article, we claim that 'an openness to other styles of work than the technocratic model has become visible, but that the introduction of the PNS [postnormal science] paradigm is still in its early stage' (Petersen et al. 2011: 363).

problem framings, expert judgements, assumed model structures, and so on. Particularly in studies of the future, for which computer simulation is often used, we must recognise our ignorance about the complex systems under study. Verification and validation of these computer models is impossible, and confirmation is inherently partial.

The challenge to scientific advisers is to be as transparent and clear as possible in their treatment of uncertainties and to be aware of the type of policy problem they are facing. Recognising this challenge, the Netherlands Environmental Assessment Agency commissioned Utrecht University to develop, together with the agency, the *Guidance for Uncertainty Assessment and Communication* (Petersen et al. 2003; Janssen et al. 2003; van der Sluijs et al. 2003, 2004, 2008). I was closely involved in the development of the guidance, first as a researcher, subsequently as project manager, and finally as programme director responsible also for the implementation of the guidance within the agency.[*] A core team was formed that worked in close consultation with other international uncertainty experts. The guidance aims to facilitate the process of dealing with uncertainties throughout the whole scientific assessment process (see Table 4.1).[†] It makes use of the typology of policy problems outlined in the previous section. It explicitly addresses institutional aspects of knowledge development and openly deals with indeterminacy, ignorance, assumptions, and value loadings. It thereby facilitates profound societal debate and negotiated management of risks. The guidance is not set up as a protocol. Instead, it provides a heuristic that encourages self-evaluative systematisation and reflexivity on pitfalls in knowledge production and use. It also provides diagnostic help regarding where uncertainty may occur and why. This can contribute to more conscious, explicit, argued, and well-documented choices.

Following a checklist approach, the guidance consists of a layered set of instruments (*Mini-Checklist*, *Quickscan*, and *Detailed Guidance*) with increasing levels of detail and sophistication (see Figure 4.2). It can be used by practitioners as a (self-)elicitation instrument or by project managers as a guiding instrument in problem framing and project design. Using the *Mini-Checklist* (Petersen et al. 2003) and *Quickscan Questionnaire* (Petersen et al. 2003), the analyst can flag key issues that need further consideration. Depending on what is flagged as salient, the analyst is referred to specific sections in a separate *Hints and Actions* document (Janssen et al. 2003) and in the *Detailed Guidance* (van der Sluijs et al. 2003). Since the number of cross references between the documents comprising the guidance is quite large, a publicly available interactive Web application has been implemented (see http://

---

[*] On 1 January 2011, in addition to my other roles, I became chief scientist of the PBL, responsible for the agency's overall system of scientific quality assurance.

[†] Only some elements of the guidance are specific to environmental assessments and, with only minor changes, the guidance can be used in any area of scientific policy advising. Furthermore, although a strong emphasis is put on assessing simulation uncertainty, the methodology encompasses all sources of information used in science-for-policy.

TABLE 4.1

Foci and Key Issues in the *Guidance*

Foci	Key Issues
Problem framing	Other problem views; interwovenness with other problems; system boundaries; role of expected results in policy process; relation to previous assessments
Involvement of stakeholders	Identifying stakeholders; their views and roles; controversies; mode of involvement
Selection of indicators	Adequate backing for selection; alternative indicators; support for selection in science, society, and politics
Appraisal of knowledge base	Quality required; bottlenecks in available knowledge and methods; impact of bottlenecks on quality of results
Mapping and assessing relevant uncertainties	Identification and prioritisation of key uncertainties; choice of methods to assess these; assessing robustness of conclusions
Reporting uncertainty information	Context of reporting; robustness and clarity of main messages; policy implications of uncertainty; balanced and consistent representation in progressive disclosure of uncertainty information; traceability and adequate backing

Sources: Petersen, A.C., Janssen, P.H.M., van der Sluijs, J.P., Risbey, J.S., and Ravetz, J.R. (2003), RIVM/MNP *Guidance for Uncertainty Assessment and Communication: Mini-Checklist and Quickscan Questionnaire*, Bilthoven: Netherlands Environmental Assessment Agency (MNP), National Institute for Public Health and the Environment (RIVM). Available at http://leidraad.pbl.nl; van der Sluijs, J.P., Petersen, A.C., Janssen, P.H.M., Risbey, J.S., and Ravetz, J.R. (2008), Exploring the quality of evidence for complex and contested policy decisions, *Environmental Research Letters* 3: 024008 (9 pp).

leidraad.pbl.nl). This Web application also offers the functionality of a prioritised to-do list of uncertainty-assessment actions and generates reports of sessions (facilitating traceability and documentation and enabling internal and external review).

In the guidance, six parts of environmental assessments are identified that have an impact on the way uncertainties are handled (see Table 4.1). These parts are

1. problem framing;
2. involvement of stakeholders (all those involved in or affected by a policy problem);
3. selection of indicators representing the policy problem;
4. appraisal of the knowledge base;
5. mapping and assessment of relevant uncertainties;
6. reporting of the uncertainty information.

A focused effort to analyse and communicate uncertainty is usually made in parts 5 and 6. However, the choices and judgements that are made in the

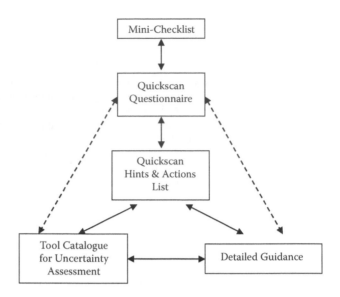

FIGURE 4.2

Components of the *Guidance for Uncertainty Assessment and Communication*. (From Petersen, A.C., Janssen, P.H.M., van der Sluijs, J.P., Risbey, J.S., and Ravetz, J.R. (2003), *RIVM/MNP Guidance for Uncertainty Assessment and Communication: Mini-Checklist and Quickscan Questionnaire*, Bilthoven: Netherlands Environmental Assessment Agency (MNP), National Institute for Public Health and the Environment (RIVM). Available at http://leidraad.pbl.nl.)

other four parts are also of high importance for dealing with uncertainty. All six parts are briefly reviewed here.

Problem framing. The problem framing part relates to the inclusion and exclusion of different viewpoints on the policy problem and the connections the policy analysis should make to other policy problems. Decisions in this part of environmental assessment influence, for instance, the choice of models (which domains should they cover, which processes should be included, etc.).

Involvement of stakeholders. The involvement of stakeholders part of environmental assessment concerns the identification of the relevant stakeholders (e.g., government; parliament; governmental advisory councils; other governmental actors at local, national, or international levels; research institutes; scientists; sector-specific stakeholders; employers' organisations; labour unions; environmental and consumer organisations; unorganised stakeholders; citizens; media; etc.) and their views on the problem, including disagreements among them. There are several ways in which stakeholders can be involved in the assessment. Either they can be involved directly or, alternatively, analysts can try to incorporate their perspectives. The guidance does not spell out how stakeholders can best be involved in environmental assessments. The present tendency within PBL is not to appoint people who represent interest groups directly in advisory panels, the democratic model of science-for-policy rejected by Jasanoff. Instead, another route is typically followed to enhance participation: to extend the system of peer review to include more diverse perspectives via experts affiliated with different governmental and societal groups. The idea is that under conditions of postnormal science, or when we are confronted with unstructured problems,

> experts whose roots and affiliations lie outside that of those involved in creating or officially regulating the issue must be brought in. These new participants, enriching the traditional peer communities and creating 'extended peer communities' are necessary for the transmission of skills and for quality assurance of results. (Funtowicz and Ravetz 1991: 149)

The definition of 'expert' is as extensive as possible. Ultimately, it is the agency responsible for the assessment that decides who count as experts and who do not. Funtowicz and Ravetz emphasise that not only is legitimacy at stake, but it is really the *quality* (defined as 'fitness for purpose') of scientific advice that stands to gain from extended peer review, especially if the system uncertainty is high:

> When problems do not have neat solutions, when the phenomena themselves are ambiguous, when all mathematical techniques are open to methodological criticism, then the debates on quality are not enhanced by the exclusion of all but the academic or official experts. (Funtowicz and Ravetz 1991: 149)

In decision making, ways must be found for modellers to engage different societal perspectives on problems. The question of what is the best way to engage these perspectives and whether nonexperts also can participate in the production of PBL assessments is still a contentious issue within PBL.* Thus, current practices within PBL do not yet live up to the guidance's ideal of real involvement of stakeholders. Still, we have witnessed recent examples of projects, such as a project on the long-term options for Dutch sustainable urban development policy, for which steps towards greater openness have been made by PBL (Petersen et al. 2011).

---

* To facilitate the internal debate on this topic and to identify tools for stakeholder participation, the Netherlands Environmental Assessment Agency commissioned the Radboud University Nijmegen to develop, together with the agency, a guidance on stakeholder participation. This guidance was finalised in 2007 (see Hage et al. 2010).

Selection of indicators. In the selection of indicators part, important choices are made with respect to output processing and interpretation: Decisions are taken on what indicators are calculated and included in the study. One should realise that alternative choices can always be made, and that sometimes alternatives are brought forward and advocated by participants in the debate. The uncertainties associated with indicators may differ depending on the indicators chosen, and indicators may be more or less representative of a problem.

Appraisal of knowledge base. The appraisal of knowledge base part involves establishing what quality of information is needed for answering the questions posed, which depends on the required quality of the answers. Bottlenecks in the knowledge and methods that are needed for the assessment may be identified, and decisions to pursue further research may be taken in the case of deficiencies.

Mapping and assessment of relevant uncertainties. In the mapping and assessment of relevant uncertainties part, the uncertainties in simulation models are characterised, and plans may be made for assessing these uncertainties more thoroughly by using standardised uncertainty assessment tools (e.g., taken from the *Tool Catalogue*; van der Sluijs et al. 2004).[*] All these activities take place with an eye to enable one to state the consequences of these uncertainties for the most policy-relevant conclusions of the study.

Reporting of uncertainty information. In the reporting of uncertainty information part, the assessors ensure that the uncertainties are adequately communicated, mainly through formulating messages that are robust with respect to these uncertainties—that is, the strength of the policy-relevant statements made is tailored to the reliability of the underlying simulation models. Some advice is given on how to communicate the different dimensions of uncertainty.[†] From the typology of uncertainty developed in Chapter 3 (Figure 3.1), it follows that there are five ways to express uncertainty:

- by *characterising the nature* of the uncertainty;
- by *presenting a range* of uncertainty[‡];

---

[*] Other tool overviews can be used as well, such as that of the Climate Change Science Program (CCSP; 2009).

[†] In preparation for the development of further guidance materials for uncertainty communication, we found empirical confirmation for the assumption in the guidance that decision makers hold a particular interest in explicit communication on the implications of uncertainty (Wardekker et al. 2008).

[‡] Note that ranges can also be expressed linguistically in a qualitative manner (cf. van Asselt 2000: 313–319).

- by *acknowledging ignorance* about the system studied;
- by *characterising the methodological quality* of the research; and
- by *acknowledging the value-ladenness* of choices.

To facilitate communication about the different types of uncertainty that arise in scientific assessments, an uncertainty typology quite similar to the typology developed in Chapter 3 of this study is part of the guidance.

In the guidance, the uncertainty matrix is also used as an instrument for generating an overview of where one expects the most important (policy-relevant) uncertainties to be located (the first dimension) and how these can be further characterised in terms of the other uncertainty dimensions mentioned. The matrix can be used as a scanning tool to identify areas in which a more elaborate uncertainty assessment is required. The different cells in the matrix are linked to available uncertainty assessment tools suitable for tackling that particular uncertainty type. These tools are described in a *Tool Catalogue* that aims to assist the analyst in choosing appropriate methods.

The *Tool Catalogue* provides practical ('how to') information on state-of-the-art quantitative and qualitative uncertainty assessment techniques, including sensitivity analysis, techniques for assessing unreliability$_2$ (Funtowicz and Ravetz 1990; van der Sluijs et al. 2005), expert elicitation, scenario analysis, model quality assistance (Risbey et al. 2005), and analysis of the value-ladenness of assumptions (Kloprogge et al. 2011). A brief description of each tool is given along with its goals, strengths and limitations, required resources, as well as guidelines for its use and warnings for typical pitfalls. It is supplemented by references to handbooks, software, example case studies, Web resources, and experts. The *Tool Catalogue* is a 'living document', available on the Web, to which new tools can be added.

The institutional challenges of implementing this new approach should not be underestimated. It entails much more than disseminating the documents through an organisation. For example, the top management of the Netherlands Environmental Assessment Agency had ordered and subsequently endorsed the guidance; the agency's methodology group led the development of the *Mini-Checklist* and *Quickscan*; the use of the guidance is now mandatory as part of the agency's quality assurance procedures, and the staff is actively trained to acquire the necessary skills. In addition, a methodological support unit is available in the agency to assist and advise in assessment projects. The required process of cultural change within the institute was consciously managed over the period since 2003. Although the guidance is not yet fully used within all projects, it is increasingly employed, and the attitude towards it has changed as participants in projects become more aware of its potential benefits. The originators of the guidance contend that the communication on uncertainty in agency reports has improved over this period as a result (van der Sluijs et al. 2008).

4.6 Conclusion

In this chapter, it was argued that by thoroughly assessing computer-simulation uncertainty and including reflection on the uncertainties in policy advice, simulation can play a meaningful role in policymaking. Thus the sound science debate, in which critics of environmental regulation showed an aversion to the use of simulation, is misguided. There are indeed significant uncertainties associated with simulations used to provide advice on complex societal problems, but these uncertainties do not automatically delegitimate all policies in which information from simulations of such policy problems has been used. Still, since policymakers are usually not themselves able to identify and assess the uncertainty of scientific simulation-model outcomes, scientific policy advisers must carefully weigh how to present their conclusions and how to communicate the uncertainties. Policy problems with high societal stakes and high scientific uncertainty attached to them pose a special challenge to scientific advisers (which I called the challenge of postnormal science). There is a need for institutions at the boundary between science and policy, boundary organisations such as the PBL Netherlands Environmental Assessment Agency and the IPCC, which can internalise procedures for dealing with the conditions of postnormal science.

I argued that the way uncertainties in simulation should be dealt with by scientific advisers depends on the type of policy problem they are confronting. The PBL made use of a typology of policy problems based on the dimensions of agreement on political objectives and consensus about relevant knowledge in its *Guidance on Uncertainty Assessment and Communication*. By improving its environmental assessment processes with respect to the framing of the problem, the involvement of stakeholders, the choice of indicators, the appraisal of the knowledge base, the assessment of uncertainties, and the reporting of uncertainties, the PBL has become better able to take on the challenge of postnormal science. In this way, more appropriate decisions can be taken on what information on uncertainties, which in environmental assessment are largely associated with simulations, needs to be included in the reports published by the PBL.

Section II

The Case of Simulating Climate Change

5

The Practice of Climate Simulation

5.1 Introduction

Climate simulations play an important role in climate science.* These simulations involve mathematical models that are implemented on computers and imitate processes in the climate system. Like the history of numerical weather prediction, the history of climate science is strongly related to the history of the computer. There are two main reasons why simulation is so important in climate science. First, computers removed a barrier in meteorological practice: The speed with which calculations could be done has been enhanced tremendously. We cannot practically do the calculations in climate simulations without the use of computers. Second, simulation is an important ingredient of climate science because real experiments with the climate as a whole are impossible. If we want to 'experimentally' manipulate climate, we need to perform such manipulations on a representation of the climate system.

It should be mentioned at the outset of this chapter that climate science is an observational science in which the scientific activities encompass much more than performing computer simulations. In fact, climate observations are pivotally important for a whole range of activities, including climate-simulation practice. From climate observations, the world's climate scientists deem warming of the earth's climate since the middle of the 19th century 'unequivocal' (Intergovernmental Panel on Climate Change [IPCC] 2007a, SPM: 5).† In 2007,

---

* *Climate* can be defined as follows (IPCC 2007a, Glossary: 942) 'Climate in a narrow sense is usually defined as the average weather, or more rigorously, as the statistical description in terms of the mean and variability of relevant quantities over a period of time ranging from months to thousands or millions of years. The classical period for averaging these variables is 30 years, as defined by the World Meteorological Organization. The relevant quantities are most often surface variables such as temperature, precipitation and wind. Climate in a wider sense is the state, including a statistical description, of the climate system'. The indicator for climate change that is most often used is global average temperature; other quantities, such as regional temperatures, precipitation, or extreme weather events, can be used as well, but the changes in these other quantities are often statistically less significant. Note that a 'statistical description' can include also more sophisticated measures than just means and variances.

† In references to IPCC reports, aside from the page number in the whole report, the part of the report is also included: for example, SPM = Summary for Policymakers; TS = Technical Summary; Ch. *x* = Chapter *x*; or Glossary.

the Fourth Assessment Report of the IPCC concluded that the global average surface temperature has increased by 0.76°C [0.57°C to 0.95°C] (90% confidence range) over the period from 1850–1899 to 2001–2005 (IPCC 2007a, SPM: 5). The uncertainty is here expressed as a range of temperature change (from 0.57°C to 0.95°C) together with the probability that the real value lies within this range (that is, 90%). For the Northern Hemisphere, it is considered likely (between 66% and 90% chance) that current temperatures are higher than all historic temperatures over the past 1,300 years (IPCC 2007a, SPM: 9).[*]

Besides temperature, precipitation is also a component of climate. Generally speaking, warmer air can hold more moisture, but changes in precipitation also depend on changes in atmospheric flow. It is considered very likely (more than 90% chance) that during the 20th century precipitation has *increased* in eastern parts of North and South America, northern Europe, and northern and central Asia but *decreased* in the Sahel, the Mediterranean, southern Africa, and parts of southern Asia (IPCC 2007a, SPM: 7). Furthermore, over most land areas in the late 20th century, it is likely, according to the climate experts, that there has been an increase in the frequency of heavy precipitation events (IPCC 2007a, SPM: 8). Such extreme events are also typically included in the description of climate.

These statements about observed climate change have been arrived at by climate scientists without the use of climate simulations.[†] This means that the sources of uncertainty are of a different kind than those encountered in simulation practice. For example, for global average surface temperature, the sources of uncertainty at the 150-year timescale are located in data and (statistical) model assumptions made in data processing: 'measurement and sampling error, and uncertainties regarding biases due to urbanisation and earlier methods of measuring SST [sea surface temperature]' (IPCC 2007a, Ch. 3: 242). For the Northern Hemisphere temperature, at the 1,300-year timescale, the sparseness of 'proxy' data[‡] (towards the past there is increasingly limited spatial coverage) is the main source of uncertainty (IPCC 2007a, SPM: 9), besides the unreliability of proxies for determining local temperatures in the past.

---

[*] Since the publication of the corresponding 'hockey stick' figure in the SPM of the Third Assessment Report (IPCC 2001), it has come under dispute (see the criticism by McIntyre and McKitrick 2005). In that figure, the results of one temperature reconstruction, by Mann et al. (1999), for the last 1,000 years are shown. Given the uncertainty ranges, other reconstructions are possible and have indeed been published. In the IPCC (2001) assessment, several different reconstructions were used, and the conclusions that are stated in the Summary for Policymakers have taken these into account. The debate has not yet subsided, however. See, for example, Visser et al. (2010) for a discussion of the 'divergence problem', which was also featured prominently in 'Climategate' at the end of 2010.

[†] Still, mathematical models—not of the type used in dynamic simulations—are involved in interpreting, interpolating, and averaging inhomogeneous proxy data (see next note).

[‡] 'Proxies' such as tree rings, corals, ice cores, and historical records are 'interpreted, using physical and biophysical principles, to represent some combination of climate-related variations back in time' (IPCC 2007a, Glossary: 951).

It is not possible, however, to deduce directly from the observations what the *causes* of the observed changes in climate are. When climate scientists want to attribute climate changes to causes or make projections into the future, they need to make use of climate simulations. One of the most important conclusions of the IPCC (2001) report was that 'most of the observed warming over the last 50 years is likely [between 66 and 90% chance] to have been due to the increase in greenhouse gas concentrations' (IPCC 2001, SPM: 10), which was followed by the even stronger conclusion in the subsequent IPCC (2007a) report that this human influence had become 'very likely [more than 90% chance]' (IPCC 2007a, SPM: 10). To arrive at these conclusions, climate simulations have been performed as a substitute for experiments. This function of simulation is crucial in climate science since there is only one historical realisation of the system under study. Real (in the sense of controlled and reproducible) experiments at the scale of the whole climate system are impossible.

The climate system can be simulated in many different ways, using more or less-detailed models of the different components of the climate system. Figure 5.1 gives a hint of the complexity of the climate system. The diagram shows the major components of the climate system (in bold), the processes and interactions that take place between them (thin arrows), and some aspects that might change (bold arrows). The climate system is best described as a 'complex system', featuring many feedbacks, both positive and negative, and nonlinearity.* In this complex global system, the processes show variations at many spatial and temporal scales. At the timescale of days, or sometimes hours, we know that the weather can vary significantly. In the context of the human-induced global-warming problem, we are more interested in changes over a timescale of decades to centuries. When one focuses on such large spatial and temporal scales, the natural temperature variations are typically very small (about 0.5°C) as compared to the variations at smaller spatial scales and timescales. However, the impacts of these natural climatic variations on ecological systems and human populations are significant, and the consequences of a likely human-caused increase that lies between 1.1°C and 6.4°C in 2100 (see Chapter 6) can be considered quite large.

Climate science, which provides the context for our discussion of climate simulation, can be regarded as a collection of diverse scientific disciplines (physics, chemistry, geology, biology) dealing with different aspects of climate. For each component of the climate system, whole interdisciplinary scientific fields have developed (e.g., meteorology for the atmosphere, oceanography for the ocean, and glaceology for sea ice, ice sheets, and glaciers; these fields, like the system, are all interconnected). For the climate system

---

* An interaction mechanism within a complex system is called a *feedback* when the result of an initial process triggers changes in a second process, which in turn influences the initial one. To give an example, if the atmosphere warms it will hold more water vapour; if water vapour increases in the atmosphere, the greenhouse effect associated with water vapour will increase, leading to an even warmer atmosphere (positive feedback). An interaction is called *nonlinear* when there is no simple proportional relation between cause and effect.

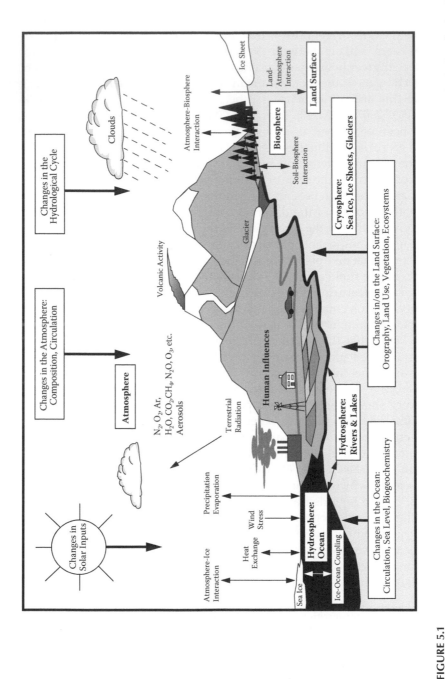

FIGURE 5.1
The global climate system. (From Intergovernmental Panel on Climate Change (2007a), *Climate Change 2007: The Physical Science Basis. Contribution of Working Group I to the Fourth Assessment Report of the Intergovernmental Panel on Climate Change,* Cambridge, England: Cambridge University Press, Ch. 1: 104.)

as it is considered in this study, human activities are treated as *input data* (of emissions and changes in land cover, leading to 'radiative forcing'[*]) in climate-simulation models. For determining the impacts of climate change on human well-being, the *output data* of climate-simulation models are used. An alternative framing of the climate system is to include human society as a dynamic component of the integrated natural–societal system and simulate human behaviour explicitly in an 'integrated assessment model' (an example is given in Chapter 7). Since the focus in this study is on simulating nature, the IPCC conceptualisation of the climate system, with humans outside its boundaries, is followed. Within the IPCC, the results of social scientific and economic studies of alternative futures for human society are specified for use in the natural scientific climate models as 'emission scenarios' (different future emission paths for greenhouse gases and aerosol[†] precursors, which depend on population, economic growth, consumption patterns, and technological developments). In 2000, for instance, the IPCC published a Special Report on Emission Scenarios (SRES), which was subsequently used in the Third and Fourth Assessment Reports of 2001 and 2007, respectively.

In this chapter, the practice of simulating the natural climate system is investigated. First, the different functions of climate simulation are investigated. Second, a comparison is made between comprehensive and simple climate models in the light of a plurality of methodological approaches to climate simulation. Third, the social context of climate simulation is addressed.

5.2 Functions of Climate Simulation

Climate simulation advanced quickly after the creation of the first three-dimensional (3-D) atmospheric general circulation models (AGCMs) in the 1950s. Since then AGCMs have increased in spatial and temporal resolution and have become increasingly complex. During this development, oceanic

---

[*] *Radiative forcing* is 'the change in the net, downward minus upward, irradiance (expressed in W m^{-2}) at the tropopause due to a change in an external driver of climate change, such as, for example, a change in the concentration of carbon dioxide or the output of the sun. Radiative forcing is computed with all tropospheric properties held fixed at their unperturbed values, and after allowing for stratospheric temperatures, if perturbed, to readjust to radiative-dynamical equilibrium. Radiative forcing is called instantaneous if no change in stratospheric temperature is accounted for' (IPCC 2007a, Glossary: 951).

[†] *Aerosols* are a 'collection of airborne solid or liquid particles, with a typical size between 0.01 and 10 μm that reside in the atmosphere for at least several hours. Aerosols may be of either natural or anthropogenic origin. Aerosols may influence climate in several ways: directly through scattering and absorbing radiation, and indirectly by acting as cloud condensation nuclei or modifying the optical properties and lifetime of clouds' (IPCC 2007a, Glossary: 941). The most important anthropogenic aerosol precursor is sulphur dioxide, SO_2, which reacts in the atmosphere to form sulphate aerosols.

processes and their interactions with the atmosphere have become included in coupled atmosphere–ocean general circulation models (AOGCMs). Currently, climate simulations are performed in a few hundred climate-simulation laboratories worldwide.[*] The most comprehensive climate models, the coupled AOGCMs, are developed and run in several tens of climate-simulation laboratories.[†]

My description of climate-simulation laboratory practice starts with an outline of the functions of climate simulation. These comprehensive climate models perform four of the five functions of simulation distinguished in Section 2.5: They can be used as a technique to investigate the detailed dynamics of the climate system; as a substitute for experiments; as a heuristic tool for developing hypotheses, models, and theories about the climate system; and as tools for observers to support climate observations. The fifth function, the educational function, cannot be fulfilled by these models due to their heavy computational demands; for that function, simplified models are used.

Firstly, the most comprehensive models are the preferred choice for the *investigation of the detailed dynamics* of the climate system. In the succinct introduction to climate models given in the Technical Summary (TS) of IPCC (2001), the structure of these comprehensive climate models is described as follows[‡]:

> Comprehensive climate models are based on physical laws represented by mathematical equations that are solved using a three-dimensional grid over the globe. For climate simulation, the major components of the climate system must be represented in sub-models (atmosphere, ocean, land surface, cryosphere and biosphere), along with the processes that go on within and between them. … Global climate models in which the atmosphere and ocean components have been coupled together are also known as Atmosphere–Ocean General Circulation Models (AOGCMs). In the atmospheric module, for example, equations are solved that describe the large-scale evolution of momentum [of atmospheric 'particles'], heat and moisture. Similar equations are solved for the ocean. Currently, the resolution of the atmospheric part of a typical model is about 250 km in the horizontal and about 1 km in the vertical above the

---

[*] This estimate is based primarily on the number of academic meteorology departments worldwide, which is about 100 (one-third of which are in the United States). In most of these departments at least simple climate models are run, for either climate research or education or both. In addition, climate simulation takes place in earth system science, oceanography, geography, geology, biology, chemistry, and physics departments. Finally, outside the universities, most developed countries have one or more research institutes in which climate simulations are being done.

[†] In the IPCC (2001) report, 34 AOGCMs from 19 modelling centres were assessed (IPCC 2001, Ch. 8), and the IPCC (2007a) report focused on 23 AOGCMS from 18 modelling centres (IPCC 2007a, Ch. 8).

[‡] Except for the fact that the resolution of the comprehensive climate models has increased over the years, this description is still valid—and can be expected to remain valid for many decades in the future.

boundary layer. The resolution of a typical ocean model is about 200 to 400 m in the vertical, with a horizontal resolution of about 125 to 250 km. Equations are typically solved for every half hour of a model integration [the time step in the model is half an hour]. Many physical processes, such as those related to clouds or ocean convection, take place on much smaller spatial scales than the model grid and therefore cannot be modelled and resolved explicitly. Their average effects are approximately included in a simple way by taking advantage of physically based relationships with the larger-scale variables. This technique is known as parametrization. (IPCC 2001, TS: 48)

A 'comprehensive' climate model is thus a model that has a relatively 'high resolution' in 3-D space and in time and that incorporates 'many processes' in 'much' detail. The most comprehensive climate models used in the IPCC (2001) report did not model many relevant biological processes, however. More recently, such processes have started to become included. Comprehensive climate models are continuously under development; more detailed elaborations of processes in the models and new processes are both regularly added to the models.

Second, comprehensive climate models are used as *substitutes for experiments* since experimentation on the climate system is impossible. For example, if one is interested in the response of the climate system to hypothetical interventions (resulting in different possible futures or counterfactual histories), this function becomes particularly prominent. The extrapolation of responses to greenhouse gas emissions leads to projections of the future response of the climate system to changes in greenhouse-gas concentrations.

Third, for some applications, such as informing policymakers of the wide range of plausible future projections of climate change associated with different emission scenarios, much simpler models than the comprehensive climate models are needed. With simple models, it becomes computationally feasible to perform many different simulations by varying model assumptions and input emission scenarios. The computational resources that would be required to perform a large number of runs with the most comprehensive models are not currently available and will not become available in the next few decades. However, use is made of comprehensive models as *heuristic tools* to construct the simple models. For instance, the parameters (e.g., 'climate sensitivity'—the equilibrium global surface temperature increase for a doubling of the equivalent CO_2 concentration) of the simple models can be determined from comprehensive models. In addition, the relationships used in simple models can be heuristically determined from comprehensive models; that is, they are not formally derived from the comprehensive models but are based on some observed patterns in the behaviour of these models.

Fourth, the comprehensive climate models are also used as *tools for observers*, for instance, in combination with satellites, to measure temperatures, cloud cover, or ozone concentrations in the atmosphere. Using climate models that include representations of processes such as clouds, atmospheric chemistry, and electromagnetic radiation one can calculate the 'spectra of electromagnetic irradiance' that will be measured by the satellite instrument under different conditions. This irradiance is dependent on many factors. First, it depends on the input from the sun (which emits mainly ultraviolet radiation). The radiation that leaves Earth lies mostly in the (long-wavelength) infrared range (while a significant amount of radiation at shorter wavelengths is reflected back into space). The outward-going infrared radiation is determined by the temperature of the earth's surface, the vertical temperature profile of the atmosphere, and the presence of clouds and greenhouse gases. The upward-going short-wavelength radiation is determined by the reflection at the earth's surface and the reflection and absorption in the atmosphere due to cloud droplets or gases that absorb long-wave radiation (such as ozone). Aside from models that describe all these radiative processes occurring in the atmosphere, simulation models of the satellite instruments themselves describe the physical transformation of the irradiance input signals into the output signals of the instrument. Such models of the instrument can be used to determine which design is optimal for the measurement of the phenomena of interest.

The fifth function of simulation, where simulation functions as a *pedagogical tool*, is not performed by AOGCMs. These comprehensive climate models are too large for this purpose. Instead, simple climate models are designed to function as interactive pedagogical tools for students or climate policymakers. The users can play with such models to get an intuitive grasp of the behaviour of the climate system by looking at the model's response to different perturbations imposed by the user. Whether such models are successful in stimulating learning depends on the quality of the graphical interface and on the ease with which the model can be run and the parameter settings and input data can be changed.

Actually, simple climate models can perform all five different functions of simulation. They can be used as a technique to investigate the detailed dynamics of a system (typically focusing on a part of the climate system); as a heuristic to develop hypotheses, models, and theories (even a simple climate model can show quite complex behaviour); as a substitute for an experiment (e.g., for future projections, as mentioned); as a tool for experimenters or observers to design experiments or measurement instruments (for a local measurement, simple models may suffice); and as a pedagogical tool to gain understanding of a process.

5.3 Varying Climate-Model Concreteness

A major methodological issue with respect to climate-simulation practice is the relationship between the simple and the comprehensive climate models.

Different climate models vary in their level of concreteness. In high-resolution, comprehensive climate models such as AOGCMs, the strategy is to resolve processes occurring within the climate system as much as possible to the smallest spatial and temporal scales of relevance. Ideally, all model relations are based on physical theory, and all empirical parameters are independently measurable. However, processes that cannot be resolved always remain, and these must be parameterised (measured against something else, i.e., against the resolved quantities). Low-resolution, simple climate models, typically one-dimensional (1-D) models, contain parameterisations at a higher level of aggregation compared to high-resolution climate models. Within the climate science community, there is a difference of opinion about the status of simple and comprehensive climate models. This methodological issue is further discussed here.

Climate models constitute one family of models in geophysical fluid dynamics. The 3-D climate system models (CSMs) are the most comprehensive models in terms of the processes that are included and relate to the largest length and timescales (the whole globe and hundreds of years). A major reason why they are different from models in geophysical fluid dynamics for the simulation of different scales and processes is that computing power is currently a bottleneck for the amount of detail these models can include. This will remain so for years to come since computing power would have to increase by several orders of magnitude to make incorporation of significantly more details possible.* The following five families of 3-D models in geophysical fluid dynamics can be distinguished (depicted in the diagram of Figure 5.2):

1. general circulation model (GCM)/CSM;
2. mesoscale model (MM), used for simulating regional weather;
3. cloud-resolving model (CRM), used for simulating storms;
4. large-eddy simulation (LES), used for simulating atmospheric/oceanic boundary layers;
5. direct numerical simulation (DNS), used for simulating small-scale turbulence.

---

* Tim Palmer observes that 'to quantify climate change with cloud-resolving climate models will require computers with substantially higher performance [than 10^{12} floating point operations per second, the maximum speed now]—we must start looking towards machines with sustained speeds in the Petaflop range (10^{15} floating point operations per second)' (Palmer 2005: 45). He claims that 'petaflop computing is not science fiction—the main high-performance computing manufacturers are actively working towards this goal and are expected to reach it in the coming few years' (Palmer 2005: 46). I am not sure whether Palmer's prediction is correct. It is definitely not certain (1) that the speed that Palmer mentions will indeed be reached within a matter of years and (2) whether the uncertainties in the models will become significantly smaller by this increase in resolution.

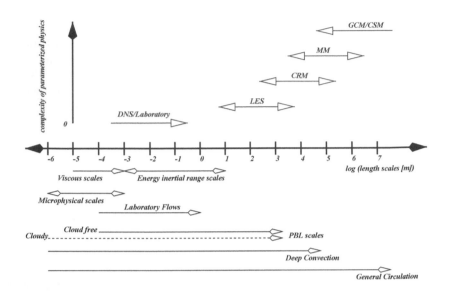

FIGURE 5.2

Families of three-dimensional models in geophysical fluid dynamics (top) and relevant physical processes (bottom). In the top part, two measures of model concreteness are used to determine the position of five families of geophysical models: the complexity of the parameterised physics (on the vertical axis) and the range of spatial scales represented by the models (roughly indicated by the length of the horizontal lines). The position along the horizontal axis signifies the physical length scales that are included in the model: A logarithmic scale is used for this horizontal axis; the axis ranges from 10^{-6} m (1 µm) to more than 10^7 m (10,000 km). In the case of DNS, no arrow is shown on the left-hand side since there is a lower bound; this lower bound is given by the smallest spatial scale at which turbulent motions occur in the atmosphere (at even smaller scales, molecular diffusion processes are dominant). Similarly, there is a largest fluid scale of relevance: the size of Earth, the upper bound for the GCM/CSM family of models. The arrows in the upper part of the figure signify that the precise domain can depend on the computational power available. In the bottom part, the different atmospheric phenomena are depicted with their ranges of relevant scales. The molecular diffusion processes dominate scales smaller than about 1 mm (*viscous scales*). In the range between about 1 mm and 10 m (the *energy inertial range*), the energy of turbulent motions is transferred from larger to smaller scales in a way that can be calculated from turbulence theory. The 'microphysical scales' are of importance in cloud-droplet processes. Since many additional degrees of freedom play a role in microphysical processes (e.g., distribution of drop sizes, chemical composition of aerosols, etc.), an arrow is drawn at the lower bound of the range of scales. The lower bound of the range of planetary boundary layer (PBL) scales is dependent on the presence of clouds: If clouds are present, microphysics plays a role. *Deep convection* is the process by which very 'active' clouds produce heavy precipitation and thunderstorms. These clouds can sometimes even penetrate into the stratosphere (in the tropics, deep convection can reach heights of about 15 km or even higher). Finally, the *general circulation* refers to global atmospheric circulation processes. (From Stevens, B., and Lenschow, D.H. (2001), Observations, experiments, and large eddy simulation, *Bulletin of the American Meteorological Society, 82*: 283–294.)

As is visible in Figure 5.2, each of these families of models spans at the most three orders of magnitude in length scales; the models are all constrained by the availability of computational resources. At the end of the previous century, there was no computational power to resolve more than about 10^9 (1 billion) grid cells. This is why the largest-scale models, the GCM/CSMs, cannot resolve horizontal scales smaller than 100 km. This is also why the smallest-scale model, the DNS, has as its own domain of application experimental laboratory flows and atmospheric flows at scales smaller than 1 m, that is, very small scales that may be of relevance in the atmosphere only for a limited range of phenomena. The grid size, or resolution, of a model determines the lower limit of the scales that can be resolved by the model.

The complete set of families of climate models (encompassing besides the 3-D models depicted in Figure 5.2 much simpler zero-, one-, and two-dimensional models) is often presented as a 'climate model hierarchy'. A consensus is presumed to exist within the climate science community for at least the last two decades about the relative merits of the different climate model families. This 'consensus view' can be found in numerous publications. The presentation here is based on the work of Harvey et al. (1997).[*] For each component of the climate system (atmosphere, oceans, terrestrial biosphere, glaciers and ice sheets, and land surface), a hierarchy of models can be identified. The main differences are in the number of spatial dimensions in the model (three, two, or one); the spatial and temporal resolution; the extent to which physical processes are explicitly represented (processes that are not explicitly represented have to be parameterised); the level of aggregation in the modelled system at which empirical parameterisations are involved; and the computational cost of running the model.[†]

According to the consensus view, both comprehensive and simple models have important roles to play in enhancing our understanding of the range of possible future climatic changes, their impacts, and interactive effects among the components of the climate system. Both pragmatic considerations involving computer resources and the level of detail appropriate to the coupling of the various components dictate the respective roles of comprehensive and simple climate models. The difference between comprehensive and simple climate models is presented by Harvey et al. (1997) in terms

---

[*] The United Nations Framework Convention on Climate Change requested that the IPCC produce a technical paper (Harvey et al. 1997) on the characteristics, strengths, and weaknesses of simple climate models in relation to more complex ones. The main reason for this request was to document the simple climate models that were used for extensive sensitivity and scenario analysis in the Working Group I volume of the 1995 IPCC Second Assessment Report. In climate science, the term *complex climate model* is often used instead of the term *comprehensive climate model*, which is preferred here since simple climate models can also include nonlinear feedbacks and thus model complexity, albeit simply.

[†] Only a partially ordered hierarchy, based on the notion of 'complexity', can be defined since the complexity of the different model aspects (resolution, number of processes included, etc.) is not necessarily correlated; some models may include many processes at low resolution or vice versa.

of the hierarchy of models introduced previously in this chapter. Typically, the behaviour of simple, 1-D climate models is easier to analyse, and sensitivity studies are easier to perform with simple models as compared with comprehensive models.

Both comprehensive and simple climate models contain empirical 'parameterisations', which are descriptions of processes not explicitly resolved in the models (e.g., convective cloud processes, which happen below the grid size) that make use of parameters that *are* available at the grid scale (e.g., temperature and humidity). All geophysical fluid dynamics models make assumptions about the influence of processes smaller than the ones resolved in the models (to the left of the length-scale ranges for models presented in Figure 5.2). One can choose either to neglect these processes, assuming they have no significant effect on the model results, or to include the net effect of these processes through parameterisations. The difference between models at varying levels of comprehensiveness is that parameterisations are introduced at different levels of aggregation. For example, in a complex, 3-D climate model the smallest scale that can be resolved for the vertical transport of heat is a few hundred kilometres (smaller-scale transport needs to be parameterised), while in a simple, 1-D climate model all vertical transport of heat by atmospheric motion is parameterised. As an aside, it is noted by Harvey et al. (1997) that very high-resolution models of clouds (LES models) have been developed with a grid spacing of tens of metres and covering several tens of square kilometres (even such models include parameterisations, e.g., parameterisations of cloud processes occurring at the micrometre scale). The ideal for some climate modellers is to extend these models to the whole globe. This is presently not possible, however, due to computing constraints, as previously mentioned.

A consequence of the fact that simple and comprehensive climate models have widely different resolutions is that in particular simple climate models (for instance, 'upwelling–diffusion climate models') the climate sensitivity and other subsystem properties must be prescribed on the basis of results from comprehensive models or observations (if available). However, in comprehensive models, such properties are explicitly calculated from a combination of resolved processes and subgrid-scale parameterisation in the models.[*] A final, qualitative difference between simple and comprehensive models is related to predictability: The 1-D simple climate models cannot simulate specific climatic 'surprises' like sudden major changes in ocean circulation, while AOGCMs can, although the timing and the nature of such changes cannot currently be reliably ascertained (amounting to recognised ignorance).[†]

---

[*] For completeness, it should be mentioned that in some other simple climate models (for instance, 'radiative–convective climate models'), the climate sensitivity is simulated, albeit relying on a very crude parameterisation.

[†] Still, simple climate models also contain nonlinearities that may lead to the modelling of other types of surprises.

Harvey et al. claim that the consensus of the climate modelling community is 'that detailed three-dimensional ... models of atmosphere and ocean dynamics, and correspondingly highly resolved models of the Earth's terrestrial and marine biota, are the long-term goals of Earth system science' (1997: 8). This statement reflects the climate science community's choice to continue the approach of adding more complexity and spatial resolution to climate models. However, the consensus in the climate science community is not as universal as is suggested by Harvey et al. (1997). Shackley et al. (1998), for instance, identify within the scientific community two strong arguments against putting all the emphasis on this approach (both arguments are related to the fact that realisation of a satisfactory comprehensive model seems distant):

1. Key parameterisations, for example, for clouds, are not fully theoretically based, and are hence often scale dependent and to some degree arbitrary. These parameterisations are crucial for climate model behaviour.

2. Models should be related to the scale of the processes involved. It has not been validated that large-scale behaviour of climate can be represented by the combined effects of smaller-scale processes that are partly resolved and partly parameterised by complex climate models.

Although Shackley et al. (1998) do not aim to discredit the comprehensive climate-modelling approach, they raise important questions concerning the methodology to be followed by climate scientists—questions that are still valid even though the specific examples may have changed.

Shackley et al.'s *first* argument makes clear that currently no climate model is of a theoretical high quality: Ad hoc assumptions are systematically involved in deriving parameterisations. Simple and comprehensive climate models differ only in respect to the specific level of aggregation at which smaller-scale processes are parameterised. Thus both types of climate simulations suffer from the methodological problem of the arbitrariness of parameterisations, that is, their ad hoc character. An example of a major ad hoc, nonphysical correction to comprehensive climate models that was widely applied in the 1990s is 'flux adjustment'—an ad hoc model fix that was introduced in coupling ocean and atmosphere GCMs (resulting in coupled AOGCMs) and that prevented long enough computer-simulation runs of an equilibrium climate. From a scientific perspective, flux adjustment is undesirable, and the practice of flux adjustment is now no longer widespread because in later versions of climate-modelling software it is no longer needed. One of the reasons why in the year 2001 (as compared to the year 1996) the IPCC could state that '[c]onfidence in the ability of models to project future climate has increased' (IPCC 2001, SPM: 19) is related to the decreased dependence on flux adjustment[*]:

> Some recent models produce satisfactory simulations of current climate without the need for non-physical adjustments of heat and water fluxes at the ocean-atmosphere interface used in earlier models. (IPCC 2001, SPM: 19)

Still, this increased confidence does not negate the fact that most climate simulations used in the first two IPCC reports (1990 and 1996) *have* made use of flux adjustment. This serves as a reminder that *all* numerical climate models include some ad hoc adjustments, even though the specifics of the adjustments may change over time. Current examples still include cloud parameterisations.[†]

As we saw in Chapter 2, some scientists hold the ideal that ad hoc corrections should ultimately be removed. According to Randall and Wielicki (1997), for instance, parameterisations are often not strictly derived from and validated with observations, and they are typically not tested for many different conditions. Furthermore, the practice of tuning, that is, adjusting the model to observations without really understanding the physical processes that are being modelled, is considered problematic by them. They argue that a parameterisation should be left unchanged when the model that includes the parameterisation is tested as a whole. Accordingly, one should refrain from tuning the parameterisations interactively to have the outcomes of the model match all available observations since this 'bad' practice cannot give one confidence in the predictive power of the model. Tuning the empirical components is allowable only if a process is very important and poorly understood. One must subsequently strive, through continuing research, to arrive at a good understanding of how to parameterise the process. If one has succeeded in that, the tuning of the model can subsequently be removed. The quality of a parameterisation depends both on its calibration with observations and on its theoretical underpinning. According to purist modellers, in particular, a parameterisation can be considered more reliable if its theoretical underpinning is strong.

There are two methodological problems related to Randall and Wielicki's outright rejection of tuning. First, their methodological proposal makes the assumption that in principle all tuning should ultimately be eliminated. However, this is not at all self-evident. Of course, the elimination of some of the adjustable parameters from the current models can lead to an

---

[*] Technically, *flux adjustment* involves the introduction of a systematic bias in the fluxes between the ocean and the atmosphere before these fluxes are imposed on the model ocean and atmosphere.

[†] In surveys performed among climate scientists in 1996 and 2003, Bray and von Storch (2007) both times found that atmospheric models were considered least able to deal with cloud processes.

improvement of these models. There is no guarantee, though, that by progressively eliminating all ad hoc corrections (even if that were possible) numerical climate models will ever acquire good predictive capacities. The Dutch meteorologist Henk Tennekes doubts whether the whole project is even possible. He observes:

> In practice a computer model always contains all sort of tricks and empirical rules, no matter how many refinements are added. The empiricism [empirical content] contained in a computer model cannot be adjusted in advance; it is tuned by repeatedly checking the performance of the model against observations, until the model finally functions in a reliable way. [Since] the climate is a one-time experiment … , the predictions of climatic models are always overtaken by the facts, regardless of how reliable the models are. (Tennekes 1994: 78–79)

In other words, the climate of the future is fundamentally unpredictable.

Second, the testing of climate models is not a straightforward affair. We have seen in Chapter 2 that verification of a model is logically impossible (Popper). Even if we use the growing observational record to test climate models, the question of their reliability still has to be dealt with when climate models are used to predict future climate change. The approach the scientific community has taken is to link the predictive capability of climate models to their performance in reproducing the historical record (both the geological record and the past period of about 150 years for which we have a global record of real-time temperature observations). If one applies Popper's (1959) philosophy about theories to models, one can claim that if the models do not fail in this regard, they should be considered 'corroborated' (corroboration is a matter of degree and depends on the severity of the tests to which the models have been put). The impossibility of establishing the absolute truth of a theory has led Popper to insist on falsifiability as the hallmark of the scientific method (see Section 2.4). Following the same line of reasoning, Randall and Wielicki (1997)—who take a model to be an embodiment of a theory, thus providing a scientific basis for predictive modelling—argue that model predictions can be proven wrong or falsified by comparison with measurements that the model was supposed to predict, and that one should strive for such falsification.

Randall and Wielicki (1997) consider falsification of the whole model (including both the 'principal hypothesis' and 'auxiliary hypotheses') possible. In Popperian fashion, this methodology takes climate models to be corroborated by each unsuccessful attempt to falsify the model. The method proposed by Randall and Wielicki (1997) is more difficult to apply for comprehensive than for simple models. One reason is that current comprehensive models involve tuning of many of the parameterisations within the model. If the model agrees with the observations against which it is tested, this could be the result of 'compensating errors' (namely, the model has been tuned as a whole). Consequently, if the model does not agree with the observations, one does not know which adjustable parameters (or even complete parameterisations) are wrong. The testing of models therefore has to be carefully framed. This is the defensible key message behind the discussion by Randall and Wielicki (1997) on the falsification of models.

Shackley et al.'s *second* argument against focusing too much on comprehensive climate models raises questions related to the complexity of climate (see also Rind 1999). It may be the case that the processes that occur across the wide range of scales modelled by comprehensive climate models can, in fact, be addressed separately for different subranges of scales. This ultimately depends on the importance of the smallest-scale nonlinearities for the large spatial and temporal scales that are considered in climate-change studies. In view of the policy context of climate simulation, a crucial question is, What are the spatial and temporal scales needed to simulate globally averaged climate change accurately at a 100-year timescale? The current answer is that we do not know—we should recognise our ignorance about this issue. We do not know the degree to which feedbacks (see note on page 97) within the climate system are influenced by the nonlinearities in the climate system and their effect on future patterns of variability (Rind 1999). In other words, we do not know whether a model of the climate system can be constructed as a hierarchy of dynamically uncoupled models (ordered by characteristic timescales) across a broad range of temporal scales (cf. Werner 1999).

The IPCC technical paper (Harvey et al. 1997) was an attempt by the climate science community to address the methodological questions of complexity through an assessment forum (although this was not the first aim of that paper). The fact that the IPCC technical paper got no further than pointing out that in the scientific climate-assessment practice pragmatic use is made of both comprehensive and simple climate models may lead to the conclusion that no universally agreed-on methodology for climate modelling exists among climate-simulation practitioners (cf. Shackley et al. 1998). The methodological analysis by Harvey et al. (1997) claims merely that climate modelling is an 'art' and that there is 'no methodological crank to turn'. I disagree with this position since even though a plurality of methodologies exists, it is possible and desirable to discuss and compare these various methodologies.[*] A nice example of such a methodological discussion within the climate modelling literature is an article by Isaac Held (2005) in the *Bulletin of American Meteorological Society*. After noting that the importance of a climate modelling hierarchy has often been emphasised, he continues:

---

[*] This is not to say that Harvey et al. (1997) do not recognise the importance of simple climate models. As in other climate-modelling 'primers' (e.g., McGuffie and Henderson-Sellers 1997), the value of simple climate models is acknowledged.

> But, despite notable exceptions in a few subfields, climate theory has not, in my opinion, been very successful at hierarchy construction. (Held 2005: 1609)

Held questions whether the comprehensive climate models really lead to improved understanding (cf. Section 2.3.4). He argues that we need simple models that capture the essential dynamics of the phenomenon being investigated in order to check whether we really understand what is happening in the comprehensive models. According to Held, 'the health of climate theory/ modeling in the coming decades is threatened by a growing gap between high-end simulations and idealized theoretical work. In order to fill this gap, research with a hierarchy of models is needed' (2005: 1614).

From the preceding discussion on the relative merits of comprehensive and simple climate models, we can infer that the plurality at the methodological level is correlated with a plurality at the axiological level (the level of aims and goals): Different climate scientists may entertain different goals of simulation in their climate-simulation practice. Can we discern such a correlation between methods and aims amongst climate-simulation laboratories? The following discussion on the sociopolitical context of climate-simulation practice addresses that question.

5.4 The Sociopolitical Context of Climate-Simulation Practice

The social and political context in which climate simulations are developed, evaluated, and applied has a significant influence on climate-simulation practice, as has been shown in the sociological work of Simon Shackley and coworkers, among other authors. Here I first summarise what Shackley et al. found (Shackley et al. 1999; Shackley 2001). They identify different styles of doing climate simulation within the research community, with the different styles embodying different standards by which to evaluate simulation models and their results.

Shackley et al. (1999) make a distinction between two styles, a 'pragmatist' one and a 'purist' one. Taking the example of 'flux adjustment', the authors show that while pragmatists consider flux adjustment to be sufficiently innocent to represent the ocean–atmosphere coupling and still arrive at meaningful results, purists 'apply seemingly more rigorous, yet still private and informal standards of model adequacy' (Shackley et al. 1999: 428). Shackley at al. are wary of flux adjustments as 'potentially covering-up model errors, influencing the model's variability, and leading to complacency in model improvements' (Shackley et al. 1999: 445). By using flux adjustment, the climate modellers were able both to present results to policymakers and to point out the need for further model development. Pragmatic and purist modellers do not usually make their tacit criteria explicit in their publications, and are the differences of opinion concerning the use of climate-simulation results for policy development are not mentioned in the IPCC reports. Why pragmatists find the application of flux adjustment acceptable for the production of policy-relevant climate runs remains hidden from view since 'perceptions of policy needs are built seamlessly into scientific interactions' (Shackley et al. 1999: 435).

A range of factors plays a role in determining both the existence of the two cultures and their membership. Shackley et al. (1999) observe that

> pragmatist and purist cultures emerge from the interplay of a hetero-
> geneous range of factors including: organisational mission, individual
> and collective research trajectories (including past work experience
> and identification of future priorities and ambitions), funding patterns,
> involvement in providing climate-impacts scientists with scenarios, the
> role of hierarchical management and/or charisma of leaders of research
> groups, and different epistemic styles. (Shackley et al. 1999: 445)

These heterogeneous factors together constitute the social and political context of climate simulation. The distinction made between the two cultures sheds light on the fundamental assumptions of different modelling approaches. Many of the pragmatists do believe that it is correct to assume the impact of flux adjustment to be small with respect to the overall results of their models. However, they have not been able to convince the purists, who believe the assumption might be incorrect and inappropriate for use. Here we encounter an uncertainty of the recognised ignorance type: In my view, at the time this controversy raged one had to remain open regarding the question of which of the schools would in the future be judged to have been right.* Not only epistemic considerations but also social and political considerations play a role in the choices that individual modellers make. The choice to employ flux adjustment in climate simulations intended to support policy advice was in my view clearly value laden.

In an investigation of the use of comprehensive models in climate-change simulations, Shackley (2001) compared two American climate-modelling groups: the Geophysical Fluid Dynamics Laboratory (GFDL, Princeton University, Princeton, NJ) and the National Center for Atmospheric Research (NCAR, Boulder, CO). The comparison was with each other and with the U.K. climate-modelling group, the Hadley Centre for Climate Predication and Research (Bracknell, England†). In that publication, Shackley identifies the following three epistemic lifestyles in the comprehensive climate-modelling community as a whole, the first two of which correspond, respectively, to the pragmatist and purist styles discussed previously:

---

* As was said in the previous section, it now seems that the pragmatists had been correct in this specific case of flux adjustment.
† The Hadley Centre moved with the MetOffice to Exeter, England, in 2003.

1. *Climate seers.* 'Those conducting model-based experiments to understand and explore the climate system, with particular emphasis on its sensitivity to changing variables and processes, especially increasing greenhouse gas concentrations' (Shackley 2001: 115). This style is similar to the pragmatist style identified by Shackley et al. (1999) and is dominant within GFDL. The specific function of climate simulation within this style—aside from being a substitute for experiments—is that of a heuristic tool to develop hypotheses about climate change.

2. *Climate-model constructors.* 'Those developing models that aim to capture the full complexity of the climate system, and that can then be used for various applications' (Shackley 2001: 115). This style compares with the purist style and is dominant within NCAR. Within this style—again in addition to being a substitute for experiments—simulation functions as a technique to investigate the detailed dynamics of the climate system.

3. *Hybrid climate-modelling/policy style.* 'The policy-influenced objectives and priorities of the research organization, as defined by its leadership, take precedence over other individual or organizational motivations and styles' (Shackley 2001: 128). The Hadley Centre is an example. One of the objectives of the centre is to perform climate simulations that can be used as input to the assessment processes by the IPCC.* Both climate seers and climate-model constructors are involved in the centre, but none of these styles dominates due to the hierarchical style of management, which enables the centre to be both policy driven and of a high scientific quality.

Shackley (2001: 129) observes that a range of factors influences which epistemic lifestyle is adopted in a climate-modelling centre. Factors that matter include disciplinary/research experience background, organisational location and objectives, main funders, main user and customers, the role of academic collaborators and users of models, the role of policymakers in negotiations over research priorities and directions, the role of organisational culture, the opportunities to treat the climate model as a 'boundary object' (e.g., between climate and numerical weather prediction research), and the role of different national cultures of research.

One of the influences on the choice of epistemic lifestyle is constituted by the political views of the climate scientists themselves on the climate-change problem. Bray and von Storch (1999, 2007) found systematic differences in climate scientists' political views at a national scale. Based on international surveys among climate scientists, they concluded that North

---

* The IPCC assessments give most weight to results that have been published in internationally peer-reviewed journals. Hence the Hadley Centre publishes its results in journals such as *Nature*.

American climate scientists perceived the need for societal and political responses to be less urgent than their German counterparts. These differences also correlate with different assessments of the quality of climate simulations. Bray and von Storch (1999) report that even though almost all climate scientists agree that the quality of climate-simulation models is limited, the U.S. scientists were less convinced of the quality of the models than their German counterparts.

Since the 'climate seers' prefer relatively simple models and the 'climate-model constructors' prefer relatively comprehensive models, the plurality of epistemic lifestyles within and among climate-simulation laboratories thus leads to different assessments of the relationship between simple and comprehensive climate models. We can conclude therefore that there is no universally agreed-on methodology for climate simulation and that different groups of climate scientists entertain different goals of climate simulation. When we assess the uncertainties in climate simulation, we should therefore pay attention to the potential value-ladenness (including sociopolitical values) of the choices made by individual modellers or modelling groups.

5.5 Evaluating the Plurality of Climate-Simulation Models

As was discussed in Chapter 2, a wide variety of methodological approaches exists for building and evaluating simulation models. In this chapter, the case for pursuing pluralism in climate modelling is argued. Wendy Parker (2006) also observes that although climate models incorporate mutually incompatible assumptions about the climate system (e.g., different physical assumptions in cloud parameterisations), they are used together as complementary resources for investigating future climate change. Indeed, climate modellers are well aware of the limitations of their models; therefore, within the context of the IPCC, 'ensembles' of models are used to assess what is happening to the climate system.

How is this to be understood philosophically? Parker observes both an 'ontic competitive pluralism' and a 'pragmatic integrative pluralism' in climate-simulation practice. Ontic competitive pluralism exists when two models make conflicting claims about the same part of the world that they are intended to model. Typically, these models are then viewed as 'competitors'. This contrasts with 'ontic compatible pluralism', which exists when there are two or more representations that can be true of the world at the same time. From the analysis in the present chapter, I conclude that the different climate models make mutually conflicting claims about how the climate system behaves. Climate scientists such as Randall and Wielicki would like to select from among the comprehensive climate models the one that actually incorporates the most realistic assumptions about the physical processes that influence climate. Parker (2006: 22) points out, however, that 'for a variety of reasons, scientists simply have been unable to identify such a model'.

How can climate models possibly be combined meaningfully in 'multi-model ensembles' if they are incompatible? According to Parker, the pluralism in climate modelling is also 'integrative', in the sense that different climate models are used together to investigate the scientific uncertainty about the climate system. In terms of the uncertainty typology introduced in Chapter 3, a scenario range of uncertainty can be determined by using several incompatible models together. Parker concludes that climate models are thus compatible not at the level of ontology but at the level of practical application. Hence her term *pragmatic integrative pluralism*, which reflects the awareness of the climate scientists that probably none of their models is correct. Thus, two types of pluralism—ontic competitive pluralism and pragmatic integrative pluralism—coexist in climate-simulation practice.

However, some climate scientists and policymakers regard the plurality of epistemic lifestyles in climate simulation as a problem. A typical example can be found in the report 'The Capacity of U.S. Climate Modelling to Support Climate Change Assessment Activities' by the Climate Research Committee of the U.S. National Research Council (NRC) (1998). The committee argues that the hybrid climate-modelling/policy style of European climate modelling centres should be copied in the United States since 'the United States lags behind other countries in its ability to model long-term climate change' (NRC 1998: 5). While '[t]he U.S. climate modeling research community is a world leader in intermediate and smaller climate modeling efforts' (NRC 1998: 1), comparatively little money has been invested in developing and running high-end comprehensive climate models.

But from a methodological point of view, the diversity in climate modelling efforts, partly reflecting differences in the social organisation of research, must be valued positively given the large uncertainties about the behaviour of the climate system.* Through the use of different models, one can obtain an initial assessment of uncertainty. Actually, there are only a few centres where the most comprehensive climate models are developed, and no 'fully satisfactory systematic bottom-up approach' (Held 2005: 1611) for developing these complex models is available. In fact,

> model builders put forward various ideas based on their wisdom and experience, as well as their idiosyncratic interests and prejudices. Model improvements are often the result of serendipity rather than systematic analysis. Generated by these informed random walks, and being evaluated with different criteria, the comprehensive climate models developed by various groups around the world evolve along distinct paths. (Held 2005: 1611)

---

* The uncertainties in climate simulation are dealt with in detail in the next chapter.

By building models of 'intermediate complexity' (Claussen et al. 2002) that are sufficiently complex to allow for the simulation of processes of interest but are easier to understand than the most complex models, it is possible to gain a better understanding of which parameterisations determine the main uncertainties of the comprehensive models. This makes more informed choices in model development possible.

5.6 Conclusion

I have shown that the roles of climate simulation in climate science are manifold. Climate models can be found to perform all the functions of simulation identified in Chapter 2. I have also illustrated that climate models of varying levels of concreteness and of varying basic assumptions exist and are valued differently by different groups of climate scientists. On the one hand, we find relatively simple climate models that do not require huge computational resources but can be used for genuine climate-scientific research. On the other hand, we encounter very comprehensive climate models that demand top-of-the-range supercomputers to work with them. For this latter category of climate models, computing power is currently a bottleneck. This situation will remain so for at least several years. The IPCC reports have taken a pragmatic stance in this matter and acknowledge that both comprehensive and simple models have important roles to play in climate science for future projection and understanding. However, in practice there still seems to be a bias towards favouring the most complex climate models, at the expense of the development of simple climate models or models of 'intermediate complexity' that can better facilitate understanding of the climate system. This should be evaluated negatively since for climate scientists to determine which direction the further development of the most complex models should take, understanding which parameterisations contribute most to the uncertainties of these models can make the development choices more informed.

I agree with Shackley's plea that to prevent one particular group of models from dominating the field, the climate-modelling community should acknowledge the 'vital and necessary role of diversity in the practice of climate science' (Shackley 2001: 131). In my view, reflection among practitioners on model plurality should be stimulated. This could help to improve the communication of uncertainties in climate simulation to policymakers and politicians (see Chapters 6 and 7).

6

Uncertainties in Climate Simulation

6.1 Introduction

What are the uncertainties involved in climate simulation? Firstly, we must be aware that it is impossible to establish the accuracy (or reliability$_1$, defined as the extent to which accurate results are obtained in a given domain—see Chapter 3) of a climate model for future prediction simply because we cannot establish this accuracy based on repeated trials. For simulations of the past, the situation is different: There is one historical realisation of Earth's climate, changing over time, with which climate-simulation outcomes can be compared. However, from determining the accuracy of a climate model for the past, we cannot derive its accuracy for the future. As an example, the paradigmatic Intergovernmental Panel on Climate Change (IPCC, 2001) diagram of simulations of the evolution of climate over the past 150 years is presented in Figure 6.1; results are shown for different climate-model inputs (external 'radiative forcings' [see page 99, first note] on the climate system, either of natural or anthropogenic origin).[*] In Figure 6.1a, the grey band denotes an ensemble of historical simulations with one climate model in which no anthropogenic emissions are included (clearly counterfactual).[†] The result is a significant mismatch between the model and the observations for the most recent decades. In Figure 6.1b, the model includes no natural external forcings from volcanoes and the sun (also clearly counterfactual). Now, the result is a significant mismatch around the middle of the century. Figure 6.1c shows the result in which both natural and anthropogenic forcings are included. The model comes close to the observations, but how reliable$_1$ is the model that produces this result? We have no way of determining this; we certainly cannot conclude from Figure 6.1c that the model is reliable$_1$

---

[*] In this chapter, the IPCC (2001) report is used for the discussion of uncertainties in climate simulation. In the IPCC (2007a) report, some uncertainties have been reduced, while other uncertainties have been enlarged or have newly emerged. I justify this choice since there is no essential difference in the types of uncertainty that existed in 2001 and 2007 and since in this book (mainly Chapter 7) I also have an historical interest in portraying and evaluating how the IPCC (2001) report reflected climate-simulation uncertainty.

[†] The width of the band is a measure for the simulated internal climate variability due to natural internal processes within the climate system (e.g., El Niño).

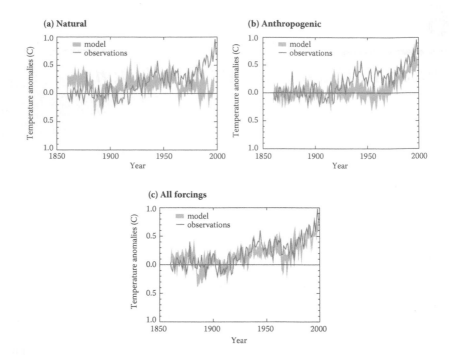

FIGURE 6.1
Comparison of model results with observations of the global mean surface temperature relative to the 1880 to 1920 mean from the instrumental record. The processes included as natural external 'forcings' (see page 99, first note) on the climate system are volcanic eruptions and changes in the solar input. The anthropogenic processes are the emissions of greenhouse gases and aerosol precursors. (From Intergovernmental Panel on Climate Change (2001), *Climate Change 2001: The Scientific Basis. Contribution of Working Group I to the Third Assessment Report of the Intergovernmental Panel on Climate Change*, Cambridge, England: Cambridge University Press, SPM: 11.)

for attributing the causes of climate change. The parameterisations in the model may be tuned to give a good fit in Figure 6.1c, and maybe some processes that are important in reality are not included in the model. The most we can do quantitatively is to apply sensitivity and uncertainty analysis to estimate—but not determine—the reliability[1] of the model.

In the IPCC (2001) report, a crude sensitivity analysis was presented by way of comparing different models (more than 30 coupled climate models were compared with each other and with observations). From this model comparison, we get a first estimate of the range of uncertainty in climate simulation. This range does not necessarily represent the full range of uncertainty about model outcomes. An assessment of the uncertainties in evaluating coupled climate models resulted in the following conclusions in the IPCC (2001) report:

> Our attempts to evaluate coupled models have been limited by the lack of a more comprehensive and systematic approach to the collection and

> analysis of model output from well co-ordinated and well designed experiments. Important gaps still remain in our ability to evaluate the natural variability of models over the last several centuries. There are gaps in the specification of the radiative forcing (especially the vertical profile) as well as gaps in proxy paleo-data necessary for the production of long time series of important variables such as surface air temperature and precipitation. (IPCC 2001 [McAvaney et al. 2001], Ch. 8: 511*)

These uncertainties affect attribution studies of the causes of climate change and are discussed in more detail in Section 6.2.

In the evaluation of climate simulations, use is made of qualitative judgements of the reliability$_2$ (the methodological rigour of the scientific procedure followed) of the simulations as well as quantitative estimates of the reliability$_1$ (the extent to which the simulation yields accurate results in a given domain). As was proposed in Chapter 3, such judgements may be based on (1) an assessment of the theoretical quality of climate models; (2) an assessment of the empirical basis of climate models; (3) comparison with other simulations; and (4) the outcome of peer review mechanisms. In the IPCC (2001) report, an entire chapter (Chapter 8, 'Model Evaluation', by McAvaney et al.) is devoted to providing information for judgements on climate-model reliability$_2$ (defined as the extent to which procedures leading to a result have methodological quality—see Chapter 3), especially qualitative judgements on relative changes in reliability$_2$ as compared with the situation in 1995, when the Second Assessment Report (SAR) of the IPCC was completed.

From an *assessment of the theoretical quality* of climate models, it is appreciated in the IPCC (2001) report that models without ad hoc flux corrections have shown improved performance over the last few years:

> Confidence in model projections is increased by the improved performance of several models that do not use flux adjustment. These models now maintain stable, multi-century simulations of surface climate that are considered to be of sufficient quality to allow their use for climate change projections. (IPCC 2001 [McAvaney et al. 2001], Ch. 8: 473)

In the previous IPCC report (SAR), the lower theoretical quality of those comprehensive models that *did* use flux adjustment was seen as a problem for those models, even though they agreed better with the observations than the models without flux adjustment. Now that models without flux adjustment show results that have come closer to the observations, the climate scientists' overall confidence in model projections has increased.

From an assessment of the empirical basis, including a partial comparison with observations, it is concluded in the report that

---

* In references to IPCC reports, aside from the page number in the whole report, the part of the report is also included: for example, SPM = Summary for Policymakers; TS = Technical Summary; Ch. *x* = Chapter *x*; or Glossary.

> [c]oupled models can provide credible simulations of both the present
> annual mean climate and the climatological seasonal cycle over broad
> continental scales for most variables of interest for climate change.
> Clouds and humidity remain sources of significant uncertainty but there
> have been incremental improvements in simulations of these quantities.
> (IPCC 2001 [McAvaney et al. 2001], Ch. 8: 473)

Thus significant differences between model simulations and observations
can be discerned with respect to clouds and water vapour. Of course, it
depends on the purpose for which a climate simulation is done how close
one judges the simulation should be to the observations to call it 'credible'.

The *replicability of model results* was addressed by comparing many models.
The uncertainties in individual models were partly averaged out using mul-
timodel ensembles in the IPCC (2001) report.

The IPCC bases its assessments on published material, preferably on
material published in peer-reviewed journals. The models included in the
IPCC (2001) model evaluation chapter had all been subjected to peer review.
However, peer review is not necessarily of high quality or necessarily objective. As has been noted, for instance, by a U.S. meteorologist, evaluation processes within meteorological research are 'currently functioning so poorly
that the integrity of the science and its timely progress are being jeopardized'
(Errico 2000: 1333). He claims that, compared to 20 years earlier, there were
more papers, presentations, and proposals; a smaller percentage of published
comments; and hardly any public discussion at conferences and workshops.
It is therefore good to note that, in addition to the peer-review mechanisms
in primary publication, the IPCC assessment process acts as a *second* peer
review mechanism (the details of which are described in Chapter 7).

The final assessment of the IPCC (2001) model-evaluation chapter is as
follows:

> Coupled models have evolved and improved significantly since the SAR.
> In general, they provide credible simulations of climate, at least down to
> sub-continental scales and over temporal scales from seasonal to decadal.
> The varying sets of strengths and weaknesses that models display lead
> us to conclude that no single model can be considered 'best' and it is
> important to utilise results from a range of coupled models. We consider
> coupled models, as a class, to be suitable tools to provide useful projec-
> tions of future climates. (IPCC 2001 [McAvaney et al. 2001], Ch. 8: 473)

The bottom line message given by the IPCC (2001) report about climate
simulation is thus that climate scientists feel confident to use these models
for climate-change studies. From the detailed evaluation of climate models,
it is possible to determine what questions can and cannot be answered and
with what degree of certainty. However, no typology of uncertainty was
used in the IPCC (2001) report to categorise the uncertainties involved in
climate simulation; similarly, no such typology was available for the IPCC

(2007a) report. In the remainder of this chapter, the IPCC (2001) report is ana-lysed with respect to climate-simulation uncertainties, making use of the typology of simulation uncertainty proposed in Chapter 3. The focus here is on comprehensive climate models. After a general discussion of uncertainty in climate simulation, the uncertainty in the causal attribution of climate change to human influences is treated in somewhat more detail.

6.2 A General Overview of Uncertainty in Climate Simulation

In Chapter 3 (Figure 3.1), a typology of uncertainty was proposed, consist-ing of six independent dimensions of uncertainty: location; nature of uncer-tainty; range of uncertainty; recognised ignorance; methodological quality; and value diversity. The six dimensions are each briefly illustrated here.

Location of climate-simulation uncertainties. In climate simulation, uncer-tainties occur at all locations distinguished in Figure 3.1: in the conceptual model; mathematical model (model structure, model parameters); model inputs; technical model implementation; pro-cessed output data and their interpretation. Some common sources of uncertainty in the conceptual model and the mathematical model structure are the following:

- *Resolved-process error.* The details of processes that are resolved on the model grid (that is, processes that involve length and tim-escales that are larger than the smallest length and timescales in the numerical model) may be erroneously modelled for different reasons. For instance, one can introduce absorption of short-wave-length radiation in resolved, that is, large-scale, clouds based on controversial measurements and without theoretical basis. Some climate models have already included a significant amount of such absorption in clouds. This may be right or wrong; additional and more reliable measurements will shed new light on this problem.

- *Unresolved-process error.* Climate models are very sensitive to parameterisations of cloud processes that take place at scales that are smaller than the model grid. The quality of all cur-rent cloud parameterisations can be considered low. Therefore, clouds constitute the most important source of uncertainty in climate models.

- *Model incompleteness.* For each application, one must ask the ques-tion whether all relevant processes have been taken into account. For instance, one cannot model the CO_2 cycle in a climate model without including an explicit representation of the biota.

Nature of climate-simulation uncertainties. We are confronted with both epistemic and ontic uncertainties in climate simulation. Climate is inherently variable—hence the importance of ontic uncertainty. Since climate-simulation models are used to determine the 'internal climate variability', we here encounter an example of epistemic uncertainty about ontic uncertainty (as discussed in Chapter 3). Within certain bounds, the surface temperature on Earth fluctuates in an unpredictable manner due to the natural variability of the climate system. This natural variability consists of two components: an 'internal variability' of the climate system (manifested in, e.g., the El Niño phenomenon) and an 'external natural variability' (related to volcanic eruptions, for instance). To determine whether, statistically speaking, the warming observed since the middle of the 20th century is due to this internal variability of the climate system or due to some other causes, such as external natural variability (volcanic eruptions, for instance) or human influences, we need to establish the magnitude of the internal variability of the climate system. Determining this ontic uncertainty requires the use of comprehensive climate models; we cannot determine it reliably from observations (since these observations are both sparse and 'contaminated' by external natural forcings). The problem now is that different models give different estimates of the ontic uncertainty—and even the range of variability as predicted by models need not include the real variability. Thus we are left with epistemic uncertainty about ontic uncertainty.

Range of climate-simulation uncertainties. The projections of future climate change that were published in the IPCC (2001) report (reproduced in Figure 6.2) show a range of plausible temperature increases by 1.4–5.8°C over the period 1990 to 2100 (Figure 6.2d). In the IPCC (2007a) report, this range was somewhat widened; the plausible range has now become 1.1–6.4°C. Part of this range is due to uncertainties in emissions (regarded as model inputs here), and part is due to climate-simulation uncertainty. The range of climate-simulation uncertainty is represented by the range of climate sensitivity simulated by different comprehensive climate models (see Section 4.3). The IPCC (2001) report treats this range as a scenario range: No probability is attached to the range of climate sensitivity of 1.5–4.5°C. More recently, modelling groups have started systematic quantification of uncertainty ranges associated with comprehensive climate models. One example is a study in which 53 versions of the Hadley Centre general circulation model (GCM) were constructed by varying model parameters (Murphy et al. 2004). Their result is a climate sensitivity range of 2.4–5.4°C, interpreted as a 5–95% probability range. A major limitation of this estimate, however, is that only

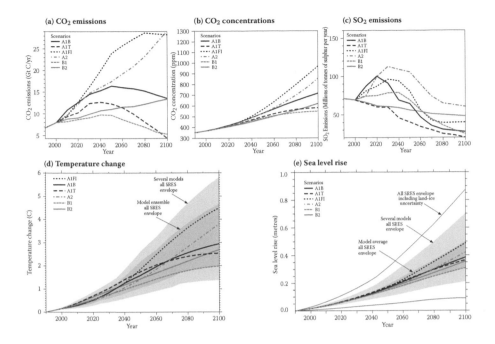

FIGURE 6.2

Uncertainty ranges for 21st century projections. The lines in (a)–(c) show the developments of CO_2 emissions, CO_2 concentrations, and SO_2 emissions for the six 'illustrative' IPCC (2000) Special Report on Emissions Scenarios (SRES). In (d) and (e), the resulting global average surface temperature change and sea-level rise are shown, as calculated with a simple climate model (which includes a small number of model parameters that are varied according to the behaviour of several comprehensive climate models). The dark grey bands show the area spanned by the full set of 35 SRES scenarios (for each scenario, the runs with varying model parameter settings have been averaged). This gives an impression of the influence of the model input uncertainty on the outcomes of interest. The light grey bands show the enhancement of the dark grey area if the models are not averaged. This gives an impression of the influence of model parameter uncertainty. (From Intergovernmental Panel on Climate Change (2001), *Climate Change 2001: The Scientific Basis. Contribution of Working Group I to the Third Assessment Report of the Intergovernmental Panel on Climate Change*, Cambridge, England: Cambridge University Press, SPM: 14.)

model parameters and not model structure has been varied. This is a more general problem in climate simulation and the assessment of its uncertainties. As van der Sluijs (1997: 173–224) shows, the best covered, but still much neglected, areas of uncertainty analysis in both comprehensive and simple climate models concern the inexactness in input data and model parameters. The other areas, which are discussed further in this chapter, receive even less attention. Still, as was said in Chapter 3, the IPCC (2007a) report, in assessing the new probabilistic estimates of climate sensitivity, arrived at the statement that it is likely (more than 66% chance) that the climate sensitivity lies in the range 2°C to 4.5°C (IPCC 2007a: 12).

Recognised ignorance in climate simulation. We must allow for the possibility of unpredictable behaviour of the climate system, which after all is a highly nonlinear system. The lack of knowledge resulting from this ontic character of the system can be classified as recognised ignorance. It is not clear at present how severe this problem is. It may be that this uncertainty will be reduced in the future, when we know more about the unpredictability of the climate system (which is a scientific subject that is being extensively studied). Still, there is some reason to believe that the climate system is relatively predictable on the timescale of 100 years. When the current comprehensive computer models of the climate system are used to calculate the nonlinear response of the climate system to perturbations, it turns out that they give approximately a linear response to increasing greenhouse gas forcing (that is, they do not show nonlinear behaviour for the perturbations used). There are more reasons to expect that climate-model uncertainties will not be so large that they render climate prediction effectively impossible. In the introductory chapter of the IPCC (2001) report, the lead authors write:

> There is evidence to suggest that climate variations on a global scale resulting from variations in external forcing are partly predictable. Examples are the mean annual cycle and short-term climate variations from individual volcanic eruptions, which models simulate well. Regularities in past climates, in particular the cyclic succession of warm and glacial periods forced by geometrical changes in the Sun-Earth orbit, are simulated by simple models with a certain degree of success. The global and continental scale aspects of human-induced climate change, as simulated by the models forced by increasing greenhouse gas concentration, are largely reproducible [by the models]. Although this is not an absolute proof, it provides evidence that such externally forced climate change may be predictable, if their forcing mechanisms are known or can be predicted. (IPCC 2001 [Baede et al. 2001], Ch. 1: 96)

Of course, these considerations do not count as proof, and we must not rest assured that the response of the climate system remains close to linear instead of becoming exponential for increased forcings. The climate system may show rapid change as a response to both internal processes or rapidly changing external forcing. Such events are considered improbable and unpredictable but not impossible, which is why IPCC reports always address the possibility of such 'unexpected events' or 'surprises'. There are many other examples of recognised ignorance in climate simulation. For instance, in Section 3.5, the example of the positive feedback between higher temperatures and higher CO_2 was introduced. The climate scientists were initially ignorant about which processes actually caused this positive feedback.

Methodological unreliability of climate simulation. Most of the qualifications of climate simulation discussed in the previous section—and indeed a large part of the discussion in this study—fall under this heading. The overall judgement by the IPCC (2001) of the methodological unreliability of climate simulation is that although each model has its strengths and weaknesses, coupled models, when taken together, are 'suitable tools' (IPCC 2001, TS: 54) for the study of climate change. According to climate scientists, models are improving and becoming more reliable₁. I give some of their reasons, following the criteria for methodological quality outlined in Section 3.6.

Theoretical basis: Although many parameterisations in climate models are still not theoretically based; some disturbing ad hoc corrections, such as flux adjustment, have been removed from some models by developing new parameterisations.

Empirical basis: Although many parts of the models have still not been rigorously compared with empirical data, the amount of available historic data and its spatial coverage are increasing and make systematic comparisons and model improvements based on these data possible. New parameterisations have been developed that are empirically more adequate.*

Comparison with other simulations: Many climate models are involved in the IPCC assessments, and intercomparison makes it possible to determine which results are replicated by the different models. Such replication does not entail that the models are correct, however, since all models may be wrong in some respects. An example of the last possibility at the time of writing of the IPCC (2001) report was the difference between climate-model estimates and measurements of the global averaged temperature of the lowest 8 km of the atmosphere over the last 20 years of the 20th century. The models showed a statistically significant warming similar to surface warming, while satellite and weather balloon measurements showed no statistically significant warming over this period.† The cause of this difference was initially unknown. Later it was argued that the measurements had been wrong: 'New analyses of balloon-borne and satellite measurements of lower- and mid-tropospheric temperature show warming rates that are similar to those of the surface temperature record' (IPCC 2007a, SPM: 5).

---

* There is a trade-off with theoretical quality here: if parameterisations are just tuned to reproduce the new data, then the theoretical quality diminishes. The best way to proceed is to try to understand where the parameterisations are going wrong and why and to propose new and better parameterisations.

† Over a longer period, from 1950, the simulation results and measurement did agree, however.

Peer consensus: The descriptions of the various comprehensive climate models are usually published in peer-reviewed journals.
The IPCC intercomparison and evaluation process provides a
second round of peer review for the different models.

However, the analysis, and especially the communication, of the
methodological dimension of uncertainty is still underdeveloped in
the IPCC, as is shown in a case study on the IPCC (2001) report in
the next chapter. This conclusion also pertains to the IPCC (2007a)
report. Much more systematic evaluations of the methodological
quality of climate simulations can and should be carried out.

Since one is confronted with epistemological and practical problems in determining the reliability$_1$ of climate models—the reliabil-
ity$_1$ of most interesting findings based on climate models cannot be
determined is my claim—I propose to shift the focus to the reliabil-
ity$_2$ of climate models. We know that models are not perfect and
never will be perfect. Models, when they do apply, will hold only in
certain circumstances. We may, however, be able to identify short-
comings of our model even within the known circumstances and
thereby increase our understanding (Smith 2002). As was observed
in Chapter 3, a major limitation of the statistical definition of reliability is that it is often not possible to establish the accuracy of the
results of a simulation or to assess quantitatively the impacts of different sources of uncertainty. Furthermore, disagreement (in distribution) between different modelling strategies would argue against
the reliability of some, if not all, of them. An alternative is therefore
to define the reliability of findings based on climate models in more
pragmatic terms. As Parker (2009) shows, to establish the 'adequacy
for purpose', or reliability, of a model in a particular context, scientists rely not only on the statistical proximity of model results to, for
instance, a historical dataset of the quantity of interest (since in that
way the reliability$_1$ for predicting the future cannot be established)
but also on a much more elaborate argumentation. This argumentation includes, I claim, an assessment of the reliability$_2$ of the model,
for instance, of the methodological quality of the representation of a
particular dynamic process that is thought to be of importance for its
use (e.g., for modelling particular future changes).

Value diversity in climate simulation. Since a range of factors influences
the assumptions made by individuals and groups of modellers in
their climate simulation practice—think of, for example, organisational mission, career paths, funding patterns, interaction with
policymakers, management and leadership styles within simulation
laboratories, and different epistemic styles—these assumptions are
potentially value laden. Examples are preferences for different styles
of climate modelling (see Section 5.4): Different climate scientists

may entertain different goals of simulation in their climate-simulation practice, and some may favour simple climate models while others favour more comprehensive ones. Furthermore, the political views of the climate scientists may influence their modelling choices. I agree with Winsberg (2010) that also model-based estimates of uncertainty about future climate change—even probabilistic estimates—are value laden.

Based on the distinctions introduced in Sections 2.5 and 3.7, the values that play a role in making climate-simulation choices are of four kinds:

General epistemic values: Scientists' general ideas about complexity—that is, whether models should be made as comprehensive as possible or kept as simple as possible—influence preferences for different modelling strategies.

Discipline-bound epistemic values: The disciplines climate scientists belong to influence the emphasis they decide to put on, for instance, physical processes versus biological processes in comprehensive climate models.

Sociopolitical values: The views of climate scientists of why they are performing their simulations (e.g., to provide policy advice) may influence choices in their models. For instance, flux adjustment was deemed necessary for some models to be able to perform simulations with political relevance.

Practical values: Practical issues such as obtaining results on time and remaining within a given budget pose limitations to climate simulation. For example, a lack of computing capacity influences the resolution of models and what processes are included in models and what processes are not.

Since a combination of values often influences the choices that climate modellers make, it is difficult in practice to determine which values in fact did influence a particular choice. Methodologies for exploring the value-ladenness of assumptions such as the one developed by Kloprogge et al. (2011) may offer some assistance in determining which values may have had more influence than others.

6.3 Climate-Simulation Uncertainty and the Causal Attribution of Temperature Change

'Detection' of climate change concerns the question of whether a warming of Earth's surface that is significantly larger than variations caused by the internal variability of the climate system can be detected. In this section, I focus on the question of whether a significant part of the detected warming could be attributed to human influences based on the state of science in the year 2000, when the IPCC (2001) report was finalised. In Figure 6.1, climate-simulation results for different radiative forcings, featuring different combinations of natural and anthropogenic forcings, are shown.

Without climate simulations, detection and attribution studies would be severely hampered:

> To detect the response to anthropogenic or natural climate forcing in observations, we require estimates of the expected space–time pattern of the response. The influences of natural and anthropogenic forcing on the observed climate can be separated only if the spatial and temporal variation of each component is known. These patterns cannot be determined from the observed instrumental record because variations due to different external forcings are superimposed on each other and on internal climate variations. Hence climate models are usually used to estimate the contribution from each factor. The models range from simpler energy balance models to the most complex coupled atmosphere–ocean general circulation models that simulate the spatial and temporal variations of many climatic parameters. (IPCC 2001 [Mitchell et al. 2001], Ch. 12: 705)

The IPCC (2001) report concluded that a warming of about 0.5°C since the middle of the 20th century could be detected, and that it was 'likely' that most of this warming was attributable to the human-induced increase in greenhouse gas concentrations. This conclusion has become one of the most important conclusions ever published by the IPCC. Where the SAR concluded in 1995 that 'the balance of evidence suggests a discernible human influence on global climate', the Third Assessment Report (TAR) conclusion gives a quantified statement ('most'), which was qualified by an assessment of the confidence experts have in this statement (they considered it 'likely' to be true, i.e., with a judgemental estimate of a 66–90% chance, as the IPCC defined *likely* in its 2001 report). The conclusion was interpreted by politicians to support their determination to implement climate policies (by reaching the Bonn Agreement in July 2001, concerning the implementation of the Kyoto Protocol of 1997). The comparison between the model simulations and observations shown in Figure 6.1 was often used to illustrate that the recent climate change can indeed be attributed to human influences. And the IPCC Fourth Assessment Report (AR4) report of 2007 went further along the line set out by the TAR by increasing the likelihood to 'very likely' (more than 90% chance).

But, what were the simulation uncertainties involved in attributing recent climate change to human influences? The five key uncertainties are described in Table 6.1. Their identification as key uncertainties and their descriptions

TABLE 6.1

Sources of Simulation Uncertainty in Climate-Change Attribution Given the
State-of-Science in 2000

Source	Short Name	Description
1	Internal climate variability	'The precise magnitude of natural internal climate variability remains uncertain. The amplitude of internal variability in the models most often used in detection studies differs by up to a factor of two from that seen in the instrumental temperature record on annual to decadal time-scales, with some models showing similar or larger variability than observed. However, the instrumental record is only marginally useful for validating model estimates of variability on the multi-decadal time-scales that are relevant for detection' (IPCC 2001 [Mitchell et al. 2001], Ch. 12: 729).
2	Natural forcing	'For all but the most recent two decades, the accuracy of the estimates of natural forcing may be limited, being based entirely on proxy data for solar irradiance and on limited surface data for volcanoes. There are some indications that solar irradiance fluctuations have indirect effects in addition to direct radiative heating, for example, due to the substantially stronger variation in the UV band and its effect on ozone, or hypothesised changes in cloud cover. These mechanisms remain particularly uncertain and currently are not incorporated in most efforts to simulate the climate effect of solar irradiance variations, as no quantitative estimates of their magnitude are currently available' (IPCC 2001 [Mitchell et al. 2001], Ch. 12: 729).
3	Anthropogenic forcing	'The major uncertainty in anthropogenic forcing arises from the indirect effect of aerosols. The global mean forcing is highly uncertain. The estimated forcing patterns vary from a predominantly Northern Hemisphere forcing similar to that due to direct aerosol effects to a more globally uniform distribution, similar but opposite in sign to that associated with changes in greenhouse gases' (IPCC 2001 [Mitchell et al. 2001], Ch. 12: 729).
4	Response patterns to natural and anthropogenic forcing	'There remains considerable uncertainty in the amplitude and pattern of the climate response to changes in radiative forcing. The large uncertainty in climate sensitivity, 1.5 to 4.5°C for a doubling of atmospheric carbon dioxide, has not been reduced since the SAR, nor is it likely to be reduced in the near future by the evidence provided by the surface temperature signal alone. ... There is greater pattern similarity between simulations of greenhouse gases alone, and of greenhouse gases and aerosols using the same model, than between simulations of the response to the same change in greenhouse gases using different models. This leads to some inconsistency in the estimation of the separate greenhouse gas and aerosol components using different models' (IPCC 2001 [Mitchell et al. 2001], Ch. 12: 729).

Continued

TABLE 6.1 (*Continued*)

Sources of Simulation Uncertainty in Climate-Change Attribution Given the
State-of-Science in 2000

Source	Short Name	Description
5	Free atmosphere trends	'There are unresolved differences between the observed and modelled temperature variations in the free atmosphere [portion of the Earth's atmosphere above the planetary boundary layer]. ... It is not clear whether this is due to model or observational error, or neglected forcings in the models' (IPCC 2001 [Mitchell et al. 2001], Ch. 12: 729).

follow from the discussion of detection and attribution uncertainties in Chapter 12 of the IPCC (2001) report (Mitchell et al. 2001).

1. *Internal climate variability.* Simulation models are used to estimate the internal climate variability for lack of a reliable direct estimate from the observational record. The bandwidth of the grey model results in Figure 6.1 represents only one model's estimate of the internal climate variability. Here only the uncertainty related to the sensitive dependence on initial conditions is taken into account. Four model runs were performed starting from different but equally probable initial conditions to determine each grey band. The resulting width of the band—or a width calculated based on model runs using different models—can be used as a statistical estimate of the internal climate variability, an ontic uncertainty.[*] Since this ontic uncertainty is exactly what one is looking for, the matrix cell (output processing and interpretation, ontic uncertainty) has been identified as a crucial uncertainty type. The statistical statement that is reached in the Summary for Policymakers of IPCC (2001) is the following: 'The warming over the past 100 years is very unlikely to be due to internal variability alone' (IPCC 2001, SPM: 10). This statement is claimed to be insensitive to the model used to estimate internal variability, and 'recent changes cannot be accounted for as pure internal variability, even if the amplitude of simulated internal variations is increased by a factor of two or perhaps more' (IPCC 2001, TS: 56). The variation of estimates of internal variability among models can be characterised as an epistemic uncertainty due to uncertainties located in the conceptual models and the corresponding mathematical models. Since we have no available statistical judgements about the quality of the different climate models used, the uncertainty range associated with the location of model structure is of the scenario uncertainty

---

[*] Even though the number of runs is only four, the runs can be compared at different times; therefore, enough data become available for a statistical estimate.

type. The value-ladenness of choices in making climate-simulation assumptions has been alluded to before. Many choices are also possible with respect to the framing of interpretations.

2. *Natural forcing.* The uncertainties in data of historical natural forcings enter the models mainly through the model inputs. With respect to the potential influence by the sun, there is recognised ignorance about possible processes—not included in climate models—through which the sun might exert a stronger influence on Earth's climate. There are many choices possible in producing the historical reconstructions. Some of the reconstructions used contain statistical uncertainty ranges; others do not. In the latter case, often different reconstructions, without statistical qualifications, are available—making it possible to express scenario uncertainty.

3. *Anthropogenic forcing.* Uncertainties in greenhouse gas emissions are located in the model inputs. The forcing by aerosols is calculated in some models as part of the simulation by using atmospheric chemistry models coupled to climate models.

4. *Response patterns.* Different models give widely differing response patterns in terms of regional temperature changes for similar forcings. In Figure 6.3, the vertical bars represent the statistical uncertainty (taking into account internal variability) about whether a computer-simulated temperature response signal can be matched with the observations. In Figure 6.3a, if a range encompasses the value of zero, this implies that the signal cannot be detected. The leftmost signal (greenhouse gas forcing only, using the Hadley Centre Coupled Model version 2—HadCM2) shows that the warming over the last century has been overestimated by a factor of two; inclusion of sulphate aerosols, which led to cooling, is thus needed to produce realistic warming trends.

5. *Free atmosphere trends.* Over the last 20 years of the 20th century, statistically significant differences between the temperature change of the lowest 8 km of the atmosphere (this is largely free atmosphere, above the planetary boundary layer), determined from satellite and weather balloon measurements, and the temperature change at the surface, had been found. This difference was not found over a longer timescale of 50 years. Climate models had not been able to reproduce the recent differences in temperature change between the surface and the lower atmosphere. If the observations were correct, there may have been something wrong in the model inputs. Alternatively, there could have been a similar structural error in all climate models. As was said in Section 6.2, in the subsequent IPCC assessment round finalised in 2007, it was concluded that the free atmosphere trends should not be considered as a source of uncertainty in climate models anymore since the earlier observations (actually, 'models of data'; cf. Section 4.2) had been proven to be wrong.

(a)

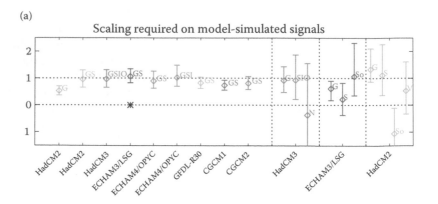

(b)

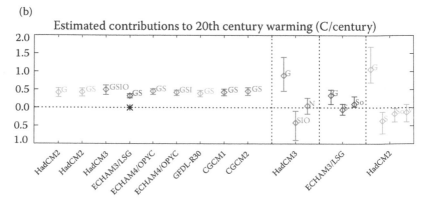

FIGURE 6.3
(a) Estimates of the 'scaling factors' by which the amplitude of several model-simulated temperature signals (G, greenhouse gases; GS, greenhouse and sulphate forcing; GSI, GS including the aerosol indirect effect; GSIO, GSI and stratospheric ozone depletion; SIO, sulphate forcing including the direct effect and stratospheric ozone depletion; N, natural external forcing, including solar and volcanic; So, solar forcing; V, volcanic forcing) must be multiplied to reproduce the corresponding changes in the observed record. The vertical bars indicate the 5–95% uncertainty range due to internal variability. Different comprehensive climate models have been used (their acronyms are written below the plots). For instance, the first model (HadCM2 with only greenhouse gas emissions and no other forcings included) overestimates the observed temperature change with a factor of two, resulting in a scaling factor of 0.5; this shows that the compensating effect of sulphate aerosols must be included in the model to obtain a more realistic simulation. (b) Estimated contributions to global mean warming over the 20th century, based on the results shown in (a). (From Intergovernmental Panel on Climate Change (2001), *Climate Change 2001: The Scientific Basis. Contribution of Working Group I to the Third Assessment Report of the Intergovernmental Panel on Climate Change*, Cambridge, England: Cambridge University Press, TS: 60.)

Figure 6.4 summarises the uncertainty sources and types that deserved particular attention in the IPCC (2001) report when the lead authors assessed and communicated climate-simulation uncertainty in climate-change attribution. The uncertainty matrix proposed in Chapter 3 is used for this purpose. The assignment of the different levels of priority is based on my own subjective expert judgement. The table contains more information than is discussed in the text. It must be regarded as an illustrative example that could trigger reflection among climate modellers and the wider scientific community. Crucial aspects to communicate are the epistemic nature of the uncertainty, the recognition of ignorance, and the methodological unreliability of the models. Since many different models are used by the IPCC, the dimension of value diversity is already partly addressed.

All the sources of uncertainty discussed impair the accuracy and reliability$_2$ of claims that can be made with respect to detection and attribution of climate change. Some of the uncertainty can be expressed statistically. However, it is only a part of the uncertainty that is captured in the statistical uncertainty statements made in Chapter 12 of the IPCC (2001) report. In addition, qualitative judgements of the reliability$_2$ of the models for the purpose of detection and attribution of climate change are indispensable. As shown in Chapter 7 of this study, the lead authors of Chapter 12 used the qualifier 'likely' instead of 'very likely' for the proposition that 'most of the observed recent warming is attributable to human influences', while the statistical model-based evidence suggests that the qualifier very likely should have been used. Although this is not explicitly stated in the IPCC (2001) report, a qualitative judgement on the unreliability$_2$ of climate models was used to argue for lowering the probability estimate to the category of likely.[*] Using climate models, a model-based inexactness estimate, and statistical procedures, the IPCC estimated that the proposition is very likely (with a 90–99% chance) to be true. The qualitative assessment of the models led the authors to change the category of the qualifier, but they did not have the vocabulary to distinguish explicitly the two uncertainty sorts of inexactness and unreliability$_2$ and their separate impacts on the main conclusion.

One of the problems with the statistical uncertainty statements in the IPCC Working Group I (WG I) reports of 2001 and 2007 is that lead authors often use the likely/very likely vocabulary in an objective probability mode (mainly in relation to observations in the past). Officially, as it also says in Footnote 7 of the IPCC (2001) SPM, the chances given in the WG I reports are 'judgmental estimates of confidence', implying a subjective probability mode. In the future, that is, in the Fifth Assessment Report (AR5), the IPCC should make transparent what kind of statistics are being used and how the assessment of the unreliability$_2$ of models is taken into account. The 'likelihood' terminology cannot adequately represent model unreliability$_2$. At

---

[*] It was stated orally by the coordinating lead author, John Mitchell, at the IPCC WG I Plenary Session in Shanghai, January 2001, and confirmed later by him in an interview.

Uncertainty Matrix / Location/Source of Uncertainty →		Sorts of Uncertainty						
		Nature of Uncertainty		Range of Uncertainty (Inexactness/Imprecision or Unreliability$_1$/Inaccuracy)		Recognised Ignorance	Methodological Unreliability (Unreliability$_2$)	Value Diversity
		Epistemic Uncertainty	Ontic Uncertainty/Indeterminacy	Statistical Uncertainty (Range + Chance)	Scenario Uncertainty (Range of 'What-If' Options)			
Conceptual model		1 2 3 4 5			1 3 4	1 2 3 4 5	1 3 4 5	1 3 4 5
Mathematical model	Model structure	1 2 3 4 5			1 3 4	1 2 3 4 5	1 3 4 5	1 3 4 5
	Model parameters	1 2 3 4 5			1 3 4	1 2 3 4 5	1 3 4 5	1 3 4 5
Model inputs (input data, input scenarios)		2 3 5		2 3	2 3	2 3 5	2 3 5	2 3 5
Technical model implementation (software and hardware implementation)								
Processed output data and their interpretation			1	1 4				1 4

The function of this matrix is to prioritise the uncertainty types that should receive priority in uncertainty assessment and communication by, for instance, the IPCC. The numbers refer to the sources of uncertainty listed in Table 6.1. The different fonts denote different levels of priority for assessing and communicating these uncertainties: 1 = important; 1 = very important; 1 = crucial. These levels have been assigned by the author.

FIGURE 6.4
Uncertainty matrix for simulation uncertainties in climate-change attribution.

least, it is difficult, if not impossible, to distill the lead authors' judgement of climate-model unreliability$_2$, as it influences the attribution conclusion, from the words *likely* or *very likely*. This was one of the factors affecting the deliberations on the attribution statement during the IPCC WG I plenary session in Shanghai in January 2001 (see Chapter 7).

6.4 Conclusion

Even though all climate models contain ad hoc 'parameterisations' and can be criticised methodologically for that reason, climate scientists generally feel confident using these models for climate-change studies. However, the IPCC lacks a typology of uncertainty that can be used to assess uncertainties more systematically. The typology of simulation uncertainty proposed in Chapter 3 can be fruitfully applied in the analysis of climate-simulation uncertainty, as was shown for the simulation-related sources of uncertainty in climate-change attribution studies. By applying the typology, it becomes immediately obvious that only part of the uncertainty can be expressed statistically. Additional qualitative judgements on the reliability$_2$ of the climate-simulation models are needed—and indeed played an important role in the production of the IPCC WG I reports. Since the vocabulary needed to distinguish explicitly between the two uncertainty sorts of inexactness and unreliability$_2$ was not available to the lead authors, the influence of their qualitative judgements on reaching their final conclusion remained largely invisible to outsiders.

7

Assessments of Climate-Simulation Uncertainty for Policy Advice

7.1 Introduction

The subject of climate change is imbued with scientific dissensus on what precisely is happening, and will happen, with the climate. Part of this dissensus is related to the large uncertainties associated with climate simulation discussed in Chapter 6. Furthermore, there is disagreement on political objectives vis-à-vis anthropogenic climate change (e.g., To what extent do we want to limit anthropogenic interference with the climate system? What should we do to mitigate the likely causes of climate change? To what extent should we prepare to adapt to it?). Perceptions of the climate-change risk vary widely both across the globe and within societies. Thus, the uncertainties are large—with climate simulation being a significant contributor to these uncertainties—and the stakes are high. This puts the problem of anthropogenic climate change in the category of 'unstructured' policy problems that are in need of a 'postnormal' problem-solving strategy (see Chapter 4).

In the climate-change debate, the stakes are indeed high. From the perspective of international relations and political science, the negotiating positions of countries within the United Nations Framework Convention on Climate Change (UNFCCC) are largely based on how they politically perceive the climate change issue, for example, as an issue that is treated equitably, that is providing a country with opportunities, or that—if not enough is done—is negatively affecting the country (van Asselt et al. 2008).

Some key players in the economy feel their existence threatened by calls for drastic reductions of CO_2 emissions. By 2030, the macroeconomic costs for multigas mitigation, consistent with emission trajectories towards stabilisation between 445 and 710 ppm CO_2-eq, are estimated at between a 3% decrease of global gross domestic product (GDP) and a small increase in GDP, compared to the baseline figure (Intergovernmental Panel on Climate Change [IPCC] 2007c, SPM:[*] 11). Over the period 2007–2030, this amounts to a

---

[*] See page 100, second note.

maximum reduction of the GDP growth rate by 0.13 percentage points/year (that is, a small but still significant fraction of the projected average yearly GDP growth rates of 2–3%). Note that some regions and sectors (obviously those involved in the oil and coal industry) will bear a particularly large share of these economic costs, while some other sectors, such as agriculture outside the tropics, will—at least initially and with some adaptation—benefit from climate change.

The stakes are also high for those who, through the projected climate change, risk damage to themselves or to things they value. For instance, some ecosystems are projected to become irreversibly damaged, species will become extinct, some developing small-island states risk disappearance with continued sea-level rise, food production may suffer in many areas, and so on (see IPCC 2007b). And finally, the stakes are high for those players who see business opportunities for more environmentally friendly technology.

Future climate policies will most likely be a mixture of adaptation by societies to human-induced climate change and mitigation of the causes of this change, mainly by reducing the emissions of greenhouse gases. We are in the tragic situation that because of increased greenhouse gas concentrations some future changes in the climate system already seem inevitable. Even if we drastically cut back our emissions and the atmospheric concentration of greenhouse gases stabilises, the global average temperature is projected to continue to rise by a few tenths of a degree per century for a century or more, while sea level is projected to continue to rise for many centuries (IPCC 2007a, SPM: 17). This is due to hysteresis in the climate system: It takes thousands of years for heat to be transported into the oceans and for a new equilibrium to be reached, and ice sheets are also likely only to respond slowly. But, the projected climate changes also pose significant risks to societies and ecosystems in the nearer future (IPCC 2007b). Already, recent regional changes in climate have been observed to affect ecosystems. Ecosystems are vulnerable to climate change, and some are projected to become irreversibly damaged. Many human social systems are sensitive to climate change, with agricultural systems especially vulnerable. Distributional aspects are important here since societies with the least resources typically have the least capacity to adapt and are most vulnerable. To summarise, since some significant climate change is already projected to be nearly inevitable, adaptation to climate change will be a necessary element of climate policy. Furthermore, if we want to avoid taking the risk that the adverse impacts become even larger, we would need to mitigate the likely cause of climate change, that is, anthropogenic sources of greenhouse gases must then be reduced, and sinks must then be enhanced.

What could the role of scientists be as policy advisers in unstructured problem contexts? In Chapter 4, I followed Hisschemöller et al. (2001) in characterising the role of science as that of 'problem recogniser'. The authority of scientists who take on this role can be assumed to reside in the scientists' ability to assess and communicate uncertainty and analyse the different values and perspectives on the problem. In practice, we find some scientists taking on this role, but also other scientists who take on any of the other roles identified in Figure 4.1. The media often feature 'advocates' who are either pro or against climate policy measures and reason back from those conclusions to their reading of climate-simulation uncertainty (read that uncertainty as either low or high) instead of being explicit about their underlying values. An example in the U.S. context is the group of esteemed physicists who founded and led the influential George C. Marshall Institute and who joined the environmental backlash movement, using the argument that most climate science is not good science (Lahsen 2008).

In the case of climate change, climate scientists are quite certain about the fact that warming has occurred over the last five decades and have judged it increasingly likely that this warming largely results from anthropogenic causes. Nevertheless, there remain important scientific uncertainties. Not only are there uncertainties concerning the specific causal processes of the observed climate change, but also there are uncertainties in future climate-change projections, uncertainties about the impacts of the projected changes, uncertainties about the costs of these impacts, and last but not least, uncertainties about the costs and benefits of different climate-change policies (adaptation and mitigation measures). These uncertainties need not be obstacles for political decision making on climate policy. This has been made explicit in 1992 by the world's nations including a formulation of the precautionary principle in the UNFCCC.* Which actions are warranted, however, is understandably a matter of intense controversy. In this situation, the assessment of uncertainties and the views that the different actors involved in the policy debate are crucial inputs for an informed political debate on the issue of climate change.

Just as people hold different views on what are acceptable risks, they also have different perspectives on uncertainties, including climate-simulation uncertainty. In the public debate on possible measures to curb CO_2 emissions, critics of the proposed policy measures typically refer to uncertainties in climate simulation. They argue that there is no empirical evidence of the problem ('we don't see *human-induced* global warming happening yet'), and that reliable prediction of climate far into the future (e.g., the year 2100) is not possible. As many of the critics currently admit that Earth's surface has warmed by about 0.5°C since the middle of the 20th century, the alleged lack of evidence is basically a negative assessment of the quality of climate simulation. After all, it is only by combining the observations in an explanatory (= theoretical) model can one attribute the observed changes to human influences rather than to natural fluctuations, as was discussed in Chapter 5. From a philosophical point of view, the critics certainly seem to have a

---

* Stephen Gardiner (2004: 577) argues 'that the endorsement by many policy makers of some form of precautionary ... approach is reasonable for climate change'. His argument is based on Rawls's criteria for the application of a maximin principle (Gardiner 2004: 577).

case. Questioning the reliability of climate simulation is certainly legitimate. Hence the uncertainties involved in climate simulation have taken on a central role in the 'sound science' debate (see Chapter 4), and to date a significant part of the political discussion on climate change has focused on the relationship of models to data (Edwards 1999). Their perspective on the issue of climate change often leads lobbyists of the coal and oil industry to take into account only the lower range of the IPCC projections of future climate change (e.g., the projections shown in Figure 6.2). They typically claim that climate sensitivity (the sensitivity of the surface temperature to a doubling of CO_2) is low.

At the other extreme, we find some lobbyists within the environmental movement who deem the upper range of the climate projections to be the most likely. Some environmentalists have criticised climate simulations for the fact that they would not be able to model abrupt changes in the climate system adequately (e.g., in the 1990s, Jeremy Leggett of Greenpeace; see Leggett 1999). From this perspective, the climate system could well be too complicated to be modelled adequately.[*]

As was said in Section 5.4, climate scientists themselves hold different political views on the climate-change problem. The way both experts and laypeople interpret climate-simulation uncertainty may be informed by the science–policy setting of the climate models. It is therefore impossible to neatly separate climate science and policy.

This chapter studies more closely how scientists may realise their role of uncertainty assessors and communicators under conditions of polarised political debate and severe scientific uncertainty, that is, in the context of an unstructured policy problem. First, the IPCC is analysed as a boundary organisation between science and politics that by virtue of its rules and ways of proceeding in the production of assessments of climate change has produced sophisticated and balanced assessments of climate-simulation uncertainty. The way the IPCC dealt with climate-change attribution uncertainty in its 2001 Working Group I (WG I) report is closely scrutinised. Second, methods of integrated assessment that explicitly include value-laden perspectives on uncertainty are evaluated with respect to their ability to inform political debate on unstructured policy problems. An example of a study by a precursor of the Netherlands Environmental Assessment Agency in the mid-1990s is examined in some detail.

---

[*] Ironically, the latter argument can also be used by the coal and oil lobby—only they would come to a different conclusion: We should not take the model outcomes seriously because nothing is the matter. In contrast, the environmental movement argues for a precautionary approach, as is also adopted in the UNFCCC.

7.2 The Intergovernmental Panel on Climate Change and Its Communication of Climate-Simulation Uncertainty

One can analyse climate science and policy as a whole in terms of a 'climate regime', being 'the suite of social, political, scientific, and economic networks and institutions (both formal and informal) that have emerged in response to human threats to the Earth's climate system' (Miller 2001: 497). In Figure 7.1, this climate regime is graphically depicted. In the middle of the diagram, one finds the IPCC. Let us here briefly recount how this climate regime, and specifically the IPCC, came about.

In the 1980s, climate scientists were very much involved in raising international political awareness for the human-induced global warming problem. This heightened awareness led to strong incentives provided by the international political community for the international scientific assessments of global warming. Meteorological and climate research—considered as a separate activity from policy advising—had already been internationalised in

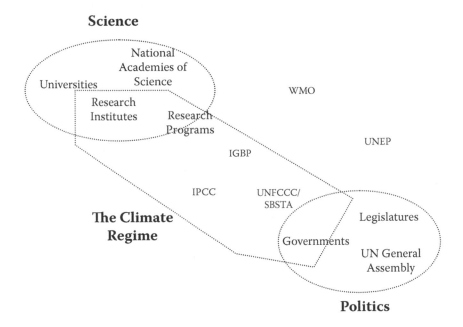

FIGURE 7.1

The institutional landscape of the climate regime. WMO, World Meteorological Organisation; UNEP, U.N. Environment Programme; IPCC, U.N. Intergovernmental Panel on Climate Change (jointly instituted by WMO and UNEP); IGBP, International Geosphere–Biosphere Programme; UNFCCC, U.N. Framework Convention on Climate Change; SBSTA, Subsidiary Body for Scientific and Technological Advice. (From Miller, C. (2001), Hybrid management: boundary organizations, science policy, and environmental governance in the climate regime, *Science, Technology, and Human Values* 26: 478–500; generalised here from U.S. institutions.)

the 1950s. Large-scale scientific cooperation through international research programs had started with the International Geophysical Year in 1957/1958. Thirty years later, in 1988, public attention for the global warming issue sharply increased in many countries (Social Learning Group 2001). That same year, at the end of the Cold War, many countries decided to cooperate on scientific climate-policy advising within the framework of the United Nations. For that purpose a new international organisation was established: the Intergovernmental Panel on Climate Change (IPCC), formally a daughter organisation of both the World Meteorological Organisation and the United Nations Environment Programme.

The IPCC can be described as a boundary organisation (see Section 4.3) between science and policy. The success of the IPCC can be measured as the degree to which this boundary organisation is able both to bring climate science to policy in a way that policymakers consider legitimate and to retain legitimacy in the scientific domain. It seems that the connection between climate science and policy has successfully been made by the IPCC. One can think, first of all, of the 1992 United Nations Framework Convention on Climate Change (UNFCCC),* the 1997 Kyoto Protocol, the 2001 Bonn Agreement, and the 2007 Bali Action Plan. Reaching those agreements was indeed facilitated by the first, second, third, and fourth IPCC assessment reports, respectively. The first report (1990) had confirmed that scientists thought that climate change may pose a serious risk, although much was still uncertain (e.g., whether the observed warming could be attributed to human influences) and had proposed ingredients for the climate-change convention. The second report (finalised in 1995) concluded that 'the balance of evidence suggests a discernible human influence on global climate' (IPCC 1996, SPM: 4), implying that the evidence for human influence had increased. The third report (2001) gave an even stronger message that 'there is new and stronger evidence that most of the warming observed over the last 50 years is attributable to human activities' (IPCC 2001, SPM: 10), and the fourth report (2007) again strengthened the message by raising the likelihood that humans have caused most of the recent warming to 90% ('very likely'). Since all governments accept the IPCC reports and approve the Summaries for Policymakers line by line, the authority of IPCC reports is acknowledged at meetings of the Framework Convention. Furthermore, the IPCC has proved to be responsive

---

* In 1992, the UNFCCC was signed in Rio de Janeiro. The assumption underlying the UNFCCC is that climate change is being caused by human activities, but the expression *climate change* had a different meaning as compared with how it is used in the context of the IPCC. To avoid misunderstanding, the IPCC repeatedly adds the following footnote to its reports: '"Climate change" in IPCC usage refers to any change in climate over time, whether due to natural variability or as a result of human activity. This usage differs from that in the United Nations Framework Convention on Climate Change, where climate change refers to a change of climate that is attributed directly or indirectly to human activity that alters the composition of the global atmosphere and that is in addition to natural climate variability observed over comparable time periods' (IPCC 2007a, SPM: 2). In this study, the expression *climate change* is used in the sense the IPCC uses it.

to requests for more tailored advice by the convention. Thus, worldwide, IPCC reports are used directly in the policymaking process. Even countries with governments that are sceptical about the Kyoto Protocol accept the full IPCC reports, albeit often reluctantly, as authoritative.

The policy relevance of the IPCC is thus ensured by its ties to the climate convention and by it being an intergovernmental body. In fact, the reason for establishing the IPCC in 1988 was the need perceived at the end of the 1980s for an international agreement on the issue of global warming (Agrawala 1998). Since some governments were not yet convinced that there was enough scientific evidence for the problem to justify actions, an intergovernmental (not just international) body was created to provide an assessment of all available knowledge on the issue that subsequently could not be discredited during the negotiation of actions. In fact, one could hypothesise that the IPCC has been so successful because the problem addressed was already considered relevant, and consensual legitimation for climate policies was precisely what was sought. This view would entail that the IPCC fulfilled a role not as problem recogniser in an unstructured problem context but as problem solver for a structured problem. However, even though some governments would like to treat the problem of climate change in this way, the large uncertainties attached to many of the findings of the IPCC and the reality of the different interests of countries in intergovernmental negotiations give rise to another hypothesis. It may have been, in particular, the assessment and communication of uncertainties and the consequent careful phrasing of the conclusions in the assessments by the IPCC that gave rise to its authority in policymaking circles.

The ties of the IPCC with political processes aimed at climate action have remained strong ever since, although the link has gradually become less direct. With the first comprehensive assessment, released in 1990, the IPCC provided direct input to the policy process. For example, the 1990 report explicitly discussed possible ingredients for a climate convention. After the convention had gained momentum in 1995, the Subsidiary Body for Scientific and Technological Advice (SBSTA) took over the discussion of matters closely related to the convention. This convention body now is the intermediary between the IPCC and the convention and has good working relations with the IPCC.

After the Third Assessment Report (TAR) was completed in 2001, the SBSTA defined its role vis-à-vis the IPCC as a forum to discuss the political impact of the IPCC conclusions in the context of the climate convention, for example, the definition of 'dangerous anthropogenic interference with the climate system' (referring to Article 2 of the UNFCCC) and, related to that, necessary future commitments (that is, emission reductions). The IPCC has made it clear that the answers to such political questions, although they must be scientifically informed, basically involve value judgements. The IPCC can only aim to provide essential information and evidence needed for decisions on what constitutes dangerous anthropogenic interference. To help the IPCC deliver the information that the climate convention needs, the SBSTA guides the IPCC in taking up policy-relevant questions, for instance by commissioning special reports or technical papers from the IPCC or by having governments submit 'policy-relevant scientific questions' to be addressed by the IPCC.

The IPCC thus tries to maintain legitimacy in the eyes of governments. Apart from the linkage to policymaking, another factor that determines the success of a boundary organisation is the degree to which the organisation is perceived by scientists to give an adequate representation of the science. In this respect, the credibility of the IPCC has been quite high since the 1990s, with a possibly temporary 'dip' in 2011 due to errors found in the regional chapters of the Fourth Assessment Report (AR4) WG II report (Meyer and Petersen, 2010; IAC 2010). IPCC reports are often used as standard works of reference for climate science (Vasileiadou et al. 2011), and the key uncertainties identified often guide priority setting for research. Still, criticism is voiced in parts of the scientific community about the direct interaction between scientists and policymakers in the production of the Summaries for Policymakers of IPCC reports. Although the number of scientists critical of the IPCC seems to have been declining over the years, some vocal critics remain. These critics usually accept the main reports as of a high scientific quality, but disqualify the Summaries for Policymakers as 'too political'. Some of these critics themselves hold the political view that climate measures should not be installed, and from their point of view the IPCC is considered to be too successful in its interaction with policymakers but unsuccessful in terms of remaining faithful to science. Since we have already concluded that concerns about the reliability of climate models are legitimate, such criticisms warrant a closer look into the assessment of simulation uncertainty by the IPCC, specifically into the writing of the Summaries for Policymakers. This investigation is taken up in the remainder of this chapter.

The strong interaction between climate science and policy, which is often publicly discussed and criticised, has led to a growing body of social science literature on this interaction, starting in the mid-1990s (a few years after the first IPCC reports had appeared). Some studies pointed out how intimately the IPCC and policymaking were actually interwoven (e.g., Boehmer-Christiansen 1994a, 1994b; Shackley and Skodvin 1995; Shackley and Wynne 1996; Skodvin 2000). This has been evaluated negatively by some social scientists as an inappropriate interaction between climate science and climate policy or more generally as a symptom of a more widespread 'problem' in environmental policymaking in which 'ambitious efforts made by scientific institutions to influence international environmental policy' have become common (Boehmer-Christiansen 1994a: 141). In this line of reasoning, the increasing attention paid by the IPCC to communicating uncertainty on the human-induced global warming problem to policymakers has led to a weakening of the influence of climate scientists on the actual policies being implemented. The UNFCCC of 1992 indeed did not include any binding emission reduction targets. And the Kyoto Protocol of 1997, with its small binding targets for the developed countries, could also be said to be much less strong than some climate scientists wanted.[*]

Still, not all analysts associate the IPCC with careful uncertainty communication, and there is indeed some evidence for deviations from the general picture that the IPCC has become more sensitive to uncertainty. The following is a quotation from former IPCC chairman Robert Watson, providing a reply at a press conference during a session of the climate convention to a reporter's question about uncertainties[†]:

> I don't think there is any doubt that the Earth's climate is changing. I don't think there is any doubt we humans are involved in that change. And I don't think there is any doubt that further climate change is inevitable, and that most of the consequences will be adverse. Are there scientific uncertainties? Yes. It is not a reason for policy inaction. (IPCC chairman, press conference, resumed session of the Sixth Conference of the Parties, Bonn, Germany, July 2001)

The second and third sentences spoken here by the IPCC chairman do not adequately reflect the uncertainty about the associated statements that is communicated in IPCC reports. He performed his role as advisory scientist very emphatically at this climate convention session. This IPCC chairman was an example of a professional serving a mediating role between science and policy, and he can be considered a member of a 'hybrid science-policy community in its own right' (Shackley and Wynne 1996: 276). The chairman in this quotation wanted to send a clear message to both the public and the politicians that we do have a global warming problem, but he did not want to offend the scientists he represented by denying the presence of uncertainties. His statements were kept vague (no numbers were given), and their impact on a lay public, including politicians, must be considered significant. However, if we take a closer look at the second and third sentences of the quotation, the chairman, by wanting to play a catalysing role at the climate convention, ran the risk of undermining the reports published by the IPCC. These do contain more cautious statements, as discussed further in this chapter.

The IPCC has often been criticised in public by scientists and lobbyists for its failure to properly communicate to policymakers the uncertainties involved in simulating climate. The quotation from the IPCC chairman at the press conference could be cited by these critics as evidence for their position. Their criticism of the chairman could run along the following lines: 'Of

---

[*] At the time of final editing for this second edition, June 2011, also the post-Kyoto process looked stuck.

[†] The quotations from meetings or press conferences presented in this chapter are transcripts of these sessions based on my notes. This quotation from Watson was obtained through the UNFCCC's webcast service and was not edited by me except for punctuation.

course there are doubts about all three propositions stated by the chairman (especially about the last two), and he implicitly acknowledges this by his allusion to the uncertainties that do remain. However, how policymakers should deal with these uncertainties is a political matter. By suppressing the communication of uncertainties in his claims that there are no doubts on the three stated propositions, he behaves irresponsibly for a scientist giving advice to policymakers, politicians, and the public. It gives them the wrong signal, namely, that the scientists are fully certain about these claims and that *therefore* the implementation of climate policies is warranted'. Similarly, critics could argue that the climate convention and the Kyoto Protocol were based on an inappropriate use of climate-change simulation. Some could go so far as to claim that projections of future climate are impeded by such formidable uncertainties that they are virtually useless (pun intended). Finally, they could criticise the use of simulations for the detection of climate change (against the background of internal climate variability) and for the attribution of detected changes to anthropogenic and natural causes, respectively. Again, their criticism could be that in climate-change detection and attribution studies the simulation uncertainties have not been properly taken into account. Such criticisms did indeed surface during the Plenary Session of WG I in Shanghai in January 2001, where the Summary for Policymakers (SPM) was approved (see Section 7.2.3).

The last sentence of the quotation from the IPCC chairman, that uncertainty is not a reason for policy inaction, borders on being policy prescriptive. IPCC procedures state that the IPCC should deliver 'policy-relevant but policy-neutral' assessments.[*] So we are confronted here with the question of how the IPCC chairman's sentence can be interpreted as being policy neutral. It is useful at this point to compare his statement to the press with what he said in his introductory talk to an IPCC internal audience of country delegates and lead authors who were present at the January 2001 meeting in Shanghai of IPCC WG I:

> The credibility of the IPCC relies on the knowledge and standing of the lead authors, reviewers and technical support units. IPCC reports are intended to be policy relevant, not policy prescribing. Governments are involved in the review process by providing comments; the challenge for the lead authors is to come up with clear and crisp answers. The IPCC has been successful: at every policy-making stage there was applause for the IPCC. We must remain diligent: the text must be absolutely correct. In Montreal [where in May 2000 the SPM of the Special Report on Land Use, Land-Use Change, and Forestry was approved] it nearly went wrong: the IPCC was very close to becoming policy prescriptive

---

[*] Since the subsequent IPCC chairman, Rajendra Pachauri, suffered from a similar problem, the IPCC has implemented, in reaction to the InterAcademy Council (2010) report, new strict principles for its communications in May 2011. These principles should prevent too political statements from being made on behalf of the IPCC.

due to attempts by some countries. The success of the IPCC was visible in The Hague: the question whether climate change is an issue was not considered relevant anymore. We have succeeded in getting across that message. (IPCC WG I cochair, plenary session, Shanghai, January 2001)

From the last sentence of this quotation it becomes clear that the chairman considers it a success for the IPCC to have influenced policymakers by having them consider climate change (used in the sense of 'human-induced climate change' here) to be an important issue. The policy neutrality of the IPCC should be taken to pertain to the contents of the measures that should be taken. Since the precautionary principle is part of the UNFCCC, the chairman's statement that the presence of uncertainties does not constitute a reason for policy inaction can be interpreted as a reiteration of the corresponding article of the UNFCCC.

Within its own proceedings, the IPCC is indeed trying to stay policy neutral. An example may illustrate this. In the first review of the TAR WG I report, the cochair of WG I was not entirely happy with the second sentence of the full report:

Many consider the prospect of human-induced climate change as a matter of concern. (IPCC 2001, Ch. 1: 87)

The cochair thought this sentence came too close to expressing concern as IPCC (Expert Comments on First Draft, January 2000). He therefore suggested referring instead to the 'possibility' of the 'prospect' of human-induced climate change and to say that because of this possibility 'it is important to ascertain the influence' of human activities. The comment was rejected by the lead authors since they did not agree that the sentence implied that the IPCC was expressing concern. The lead authors merely wanted to note that there was concern, which seems to be correct. To some readers, the sentence might not look that innocent, however, especially when read together with the sentence that follows:

The IPCC Second Assessment Report presented scientific evidence that human activities may already be influencing the climate. (IPCC 2001, Ch. 1: 87)

It can be interpreted to mean that the concern is justified by the evidence provided by the IPCC. This justificatory step is not explicitly taken here by the IPCC but is left to politicians and the public to interpret. Of course, since the IPCC indeed presented preliminary evidence of a human-induced global warming problem, and since the effects of the future warming were projected to be largely adverse, one can easily argue that concern is justified.

When we evaluate the communication of uncertainties by the IPCC, we have to pay attention to the fact that the IPCC has some influence on future funding for climate science.* The climate convention in its setup is already very research intensive, as opposed to alternative setups that can be imagined, in which strong precautionary action could be taken without requiring additional scientific advice. In this situation, climate scientists may have an interest in communicating through the IPCC mainly uncertainties instead of very strong messages without caveats. If we indeed interpret the IPCC scientists' intentions as pursuing only their own financial interests, their efforts in uncertainty communication can be evaluated as follows:

> The primary interest of research is the creation of concern in order to demonstrate policy relevance and attract funding. (Boehmer-Christiansen 1994a: 141)

However, this somewhat negative reading of climate scientists' intentions is implausible as a single explanation for increased uncertainty communication. A better explanation for the increasing attention being paid to uncertainties by the IPCC is given by procedural shifts and changed participation in the production of the IPCC assessments (Shackley and Skodvin 1995). More generally, the role of procedures, especially those concerning the review of IPCC reports, is important in structuring the science–policy interaction that takes place through the IPCC (Skodvin 2000). The IPCC is a boundary organisation that was specifically designed for the purpose of this interaction and has subsequently evolved in practice to further improve the structuring of the interaction. Of course, the explanation of the increasing attention being paid to uncertainty that focuses on the subtle interplay of actors mediated by procedures does not entail that the scientists have no interest in communicating uncertainty, but that this interest should not be regarded as the driving force.

In the IPCC process, political and epistemic motives can be found to be intertwined, sometimes leading to the suppression of uncertainty communication. The IPCC process is inevitably a politicised one due to the formal ties to the climate convention, but since in the IPCC proceedings one tries to adhere to rules of procedure, the number of times the politicisation is allowed to surface is minimised. Everyone involved in the IPCC has implicit ideas about the impact of IPCC statements on the policy process. These ideas are sometimes made explicit, as Shackley et al. (1999) experienced when they received hostile reactions from some of the pragmatist modellers involved in their attempt to study differences of opinion on climate simulation in the climate-science community.† These pragmatists thought that publications on flux adjustment would diminish the effectiveness of climate policy. I have experienced similarly hostile reactions from a reviewer of the *Bulletin of the*

---

* IPCC WG I assesses the state of climate science as expressed in the scientific literature. The IPCC does not itself perform or fund research.
† Such reactions are understandable: Shackley et al. as social students of science are rightfully considered by the modellers as actors in the societal debate on climate simulation.

American Meteorological Society, who commented on a paper of mine* about the undervalued role of simple models as compared to comprehensive models (e.g., general circulation models, GCMs) that outsiders should not meddle with climate scientists' internal affairs. Shackley et al. (1999) hypothesised that the public expression of 'intrapeer community differences' was subdued due to the presence of greenhouse sceptics in society, who are typically very vocal critics of the IPCC. Shackley et al. ironically observed that there seemed to be an agreement between some of the IPCC lead authors and the sceptics on the political consequences of putting more emphasis on uncertainties in the summaries of the reports. They advised the IPCC to accept politicisation as a given and 'to find ways to communicate informed agreement and disagreement, and informed judgements concerning levels of confidence in knowledge claims, as well as divulging the processes by which assumptions are formed and disagreements resolved' (Shackley et al. 1999: 448). The solution suggested by Shackley and coauthors was that the scientists involved should abandon the idea that communicating uncertainties inevitably leads to disbelief and policy inaction. Of course, in reality uncertainties are often politicised, but the ideal that I, with Shackley et al., wish to uphold is that the different perspectives on uncertainty can be made more explicit and can themselves become part of societal debate.

It appears that through regular revisions of both the scope of the reports and the rules of procedure, the IPCC has adjusted to external criticism, most recently in May and November 2011.† Many social scientists have published negative evaluations of how the early IPCC had treated critics of both the scientific claims and the policy proposals put forward by the IPCC (as mentioned, until 1990 the IPCC had the task of making policy proposals; from 1990 onwards, this task was taken over by other bodies).‡ Furthermore, some scientists criticise the IPCC for allowing direct interaction between scientists and policymakers in the production of the Summaries for Policymakers of IPCC reports. To be successful as a boundary organisation, however, such an interaction is definitely needed. The boundary between science and politics clearly needs continuous maintenance. As Guston writes:

---

* Petersen (2000c).
† In the November 2011 Plenary of the IPCC I myself acted as rapporteur for the task group on procedures.
‡ Social scientists have also been vocal critics of the domination of the early IPCC by natural scientists. For the first assessment report of 1990, this natural science bias can readily be demonstrated: Social scientists participated in its production only on a small scale. This changed, however, in the next round of comprehensive assessment: One of the three parts of the Second Assessment Report (SAR), which appeared in 1996, was produced by the working group Economic and Social Dimensions of Climate Change, and many social scientists participated in its production. While the SAR was under production, a broad community of social scientists decided to produce an independent extensive review of social science literature of relevance to the human-induced global warming problem (Rayner and Malone 1998). Much of the material of that review can be found again in the TAR, which like all IPCC reports provides a 'synthesis of material drawn from available literature' (IPCC Procedures).

> The success of a boundary organization is determined by principals [a term from principal-agent theory] on either side of the boundary. ... The success of the organization in performing [the] tasks [of pleasing both sides] can be taken as the stability of the boundary, while in practice the boundary continues to be negotiated at the lowest level and the greatest nuance within the confines of the organization. (2001: 401)

The question then becomes what safeguards have been built into the IPCC procedures (both formal and informal) for retaining a certain level of 'stability of the boundary'. In the remainder of this section, first the IPCC review process is analysed (Section 7.2.1) and sceptical criticism of this process, as it occurred in the TAR, is investigated (Section 7.2.2). Subsequently, the negotiations 'at the lowest level and the greatest nuance' are pictured and interpreted with respect to the SPM formulation concerning the likelihood of human-induced warming (Section 7.2.3). The purpose of the last analysis is to study closely a crucial aspect of the final report, related to problematic aspects of the uncertainty vocabulary of the IPCC TAR and AR4 assessment reports and its impact on the communication of climate-simulation uncertainty. Finally, the guidance materials that have been prepared by the IPCC, including the latest material for AR5, to assist the lead authors in their assessment and communication of uncertainty are evaluated (Section 7.2.4).

7.2.1 The IPCC Review Process

The IPCC has always paid a significant amount of attention to the quality of its review process. Compared to the traditional peer review process for journal articles, the peer review process for IPCC reports is vastly larger in scale and much more sophisticated in procedure. Some numbers related to the WG I contribution to the TAR, titled *Climate Change 2001: The Scientific Basis* (IPCC 2001), may give an impression of the amount of work involved in the production of IPCC reports.* The 14 chapters of the TAR WG I report were written by 122 lead authors and 515 contributing authors, who had started their writing in July 1998. One and a half years later, in January 2001, when the final versions of the chapters were accepted at the IPCC WG I plenary

---

* The procedures and actual way of proceeding for the AR4, finalised in 2007, were identical as compared with the TAR; only the scale was larger. This claim partially derives from my personal observation; I attended both the TAR and AR4 plenaries (respectively in Shanghai and Paris) in which the WG I SPMs were approved. Since the specific case studied in this chapter refers to the TAR, the dates and numbers given here pertain to that report.

session in Shanghai, four revisions had been made of drafts of the chapters.* The review rounds involved 420 experts and 100 governments. At the plenary session in Shanghai, the SPM of the report was approved line by line by the governments in four days. The approval of the SPM went hand in hand with the final revision of the chapters: Where the final wordings of the SPM differed from the text contained within the chapters, the precise wording of the chapters was revised accordingly for reasons of consistency.

The review comments on both the chapters and the SPM were forwarded to the lead authors, who came together for lead author meetings, consisting of lead author plenaries and chapter meetings. The lead authors were asked to write down explicitly what was done with each comment received. It is important to note here that, since the quality of the review process is not guaranteed by simply involving a large quantity of people, one has to look at the quality of the review comments that were submitted. In fact, many of the comments turned out to be of a quality similar to good article review comments.

Review editors then checked whether all review comments had received fair treatment. An important task for the review editors was to guide the lead authors in their treatment of genuine scientific controversies. The role of review editor was newly added in the IPCC procedures after the SAR. In the first and second assessment reports, a similar role was played by the working group bureaux (consisting of elected officials, mostly government scientists, who manage the working groups) and technical support units (TSUs, consisting of staff members assisting the production of working group reports). After the completion of the SAR, commentators had observed that more explicit rules of procedure were needed, while recognising that the IPCC should not become a 'science-stifling, inflexible bureaucracy' (Edwards and Schneider 2001: 228). There is a tension between scientific informality and the adherence to formal rules of procedure:

> One of the IPCC's most important features is its openness and inclusivity; balancing this against scientific informality will require constant vigilance, and perhaps a reconsideration of the formal review process. (Edwards and Schneider 2001: 228)

Through the new procedural rules, the editorial role was explicitly defined, enhancing the transparency of the review process. Review editors were asked before the plenary sessions whether they had ensured that 'all substantive expert and government review comments' had been 'afforded appropriate consideration', and that 'genuine controversies' had been 'reflected

---

* The following four official drafts were prepared: First Draft (expert review; November 1999), Second Draft (government and expert review; April 2000, including the first draft of the SPM), Final Draft (government review; October 2000), and Shanghai Draft (last-minute draft of the SPM taking the comments on the Final Draft into account; prepared for use at the plenary session in Shanghai; January 2001).

adequately in the text of the Report' (IPCC Procedures). For all 14 chapters
of the WG I report, all review editors (two per chapter) responded positively
to these questions. A further innovation that increased the transparency of
the process was the possibility for all participating reviewers to obtain all
review comments and the comments by the lead authors on these comments
through e-mail from the TSUs.*

Although the editors of the TAR WG I report had hoped that the report
would become less voluminous than the SAR WG I report (which had contained 572 pages), the authors did not succeed in keeping it short. The whole
report became 944 pages. This happened despite the fact that the report's
chapters, following IPCC's TAR Decision Paper of 1997, primarily assessed
information published since 1995, the year that the SAR had been finalised.
The 14 chapters comprise most of the 700 pages or so of the TAR WG I report.
The growth in volume of the chapters was primarily related to the sheer
increase in the number of scientific publications dealing with the issue of
climate change. All the information contained in the individual chapters
(which each have an Executive Summary) was summarised into one SPM of
17 pages (i.e., about 2% of the total volume occupied by the chapters).

Given the politicisation of the global warming issue, it is understandable
that much of the criticism of the IPCC has been directed at the SPMs, specifically at the way these are reviewed at final plenary sessions, where governments have to approve the SPM text, tables, and figures in detail, that is, line
by line. In principle, IPCC plenary sessions operate by consensus. Therefore,
everything is done to ensure that all governments can agree with the SPM.
Since governments have different political agendas, they also hold different views on what constitutes a proper 'balance' (a word used often during
plenary sessions) between the amount of space devoted to positive claims
(concerning what we know about climate change) and the amount of space
devoted to negative claims (concerning the uncertainties that remain). Given
this context, it is interesting to see how one of the cochairs of WG I introduced the governmental approval process in Shanghai (each working group
has two cochairs: one from a developed and one from a developing country;
the WG I cochairs were Sir John Houghton from the United Kingdom and
Prof. Ding Yihui from China):

> The IPCC provides a scientific assessment; therefore all proposals for
> changes in the SPM must be related to scientific accuracy, scientific bal-
> ance, clarity of message, understandability to policy makers and rele-
> vance to policy. The procedure is—based on the October text [Final Draft,
> October 2000]—to proceed bullet by bullet or sentence by sentence. The
> proposals for change by the lead authors, in response to government
> comments, are then considered. New proposals for wording changes can

* The role of review editors will be strengthened in the AR5. Also the review comments will be
made publicly available online immediately after the publication of the report. For AR4, the
review comments were posted online several years after the AR4 had been published.

> be made by the delegates. These proposals are checked with the lead
> authors for accuracy, balance, and consistency with the chapters. If pos-
> sible, the plenary should reach agreement on the new text, otherwise
> the text will be referred to either a small group to construct new draft
> wording among agreed lines, or to an open contact group to work with
> the lead authors to resolve issues and construct a new draft text. If the
> agreed SPM text implies changes in the technical summary or the chap-
> ters, lead authors will make the necessary changes and present these to
> the plenary towards the end of the meeting. (IPCC WG I cochair, plenary
> session, Shanghai, January 2001)

Thus the cochair made clear which five normative criteria were allowed to
play an explicit role during the meeting. Any proposal for changes in the
text—also if they were politically motivated—should thus be cast in terms of
'scientific accuracy', 'scientific balance', 'clarity of message', 'understandability
to policymakers' and 'relevance to policy'. Furthermore, the important role
of the lead authors came to the fore: Although it is formally the governments
that decide on the text, they are not free to make whatever changes they want.*

After this introduction by the cochair, one government raised its flag
and was given the floor. This government expressed its particular concern
that the lead authors would have too much influence on the final text by
being allowed to apply criteria such as balance themselves. According to
this country, the governments were responsible for the text, and not the lead
authors. Furthermore, the country was afraid the plenary in practice would
not discuss the Final Draft (October 2000) but would instead discuss the
new 'Shanghai Draft' prepared by the lead authors just before the meeting.
Since countries had submitted their comments based on the Final Draft and
had prepared to work with those comments, it would be too difficult for the
countries to work with a new draft that, on the one hand, most countries had
not yet read and, on the other hand, contained quite substantial changes.
The cochair at this point tried to steer the meeting away from politicisation:

> This is a scientific meeting, consisting of a scientific debate, where, of
> course, governments should decide on their positions. The lead authors
> are here to help us. The starting point shall be the October text, which
> will be projected on the screen; the Shanghai Draft is only intended to be
> of help. Regarding the criterion of balance, it is a scientific balance that
> should be strived for, not a political balance. (IPCC WG I cochair, plenary
> session, Shanghai, January 2001)

Through such rituals, which are part and parcel of most IPCC plenary sessions, the criteria regulating the changes that can be made to the text are made
explicit and thereby given extra force. Later during the meeting, references

---

* In May 2011, the important role of the lead author to check the consistency of proposed
changes with the underlying chapters—a role that was already performed over the past two
decades—was given a more formal basis in the IPCC procedures.

are often made to the criteria mentioned at the beginning of the session. Actually, in practice, the plenary session did not use the Final Draft instead of the Shanghai Draft, even though it had agreed to do so. Apparently most countries agreed with the lead authors that it was more efficient to start the discussion from the latest version produced by the lead authors. The country that had first made the objection preferred not to push the issue.

Since the aim of the IPCC is to produce reports that are credible not only among scientists and governments but also within society at large, representatives of nongovernmental organisations are admitted to IPCC sessions as observers, and experts from all organisations that represent interested or affected parties are invited to participate in the review process. The TSUs have considerable freedom to circulate the drafts widely for review. The following experts are eligible (and actively approached) to participate (IPCC Procedures)*:

- Experts who have significant expertise or publications in particular areas covered by the report.
- Experts nominated by governments as coordinating lead authors, lead authors, contributing authors, or expert reviewers as included in lists maintained by the IPCC Secretariat.
- Expert reviewers nominated by appropriate organisations.

It is the 'appropriate organisations' category that makes it possible for TSUs to really open up the IPCC review process. The WG I TSU for the TAR considered this category to include at least every organisation that expressed an interest. For instance, in the TAR WG I review comments one can find comments from special interest organisations (including fossil fuel lobbies and environmental organisations). Some of the stakeholders, notably those representing the interests of fossil fuel industries and oil-exporting countries (but also several independent sceptics, typically asked to be involved for their expertise), have repeatedly claimed that their views were not seriously considered in the IPCC reports. It is true that special interest organisations do not co-decide on the text, and in general observers are not even allowed to speak at the plenary sessions. However, their viewpoints, as expressed through the expert review rounds, are seriously considered by the authors, and through the review editor mechanisms checks are included of the way lead authors handle their comments.

The IPCC review process adds another layer to the traditional peer review that takes place in scientific practice. As was discussed in Chapter 3, peer review is a necessary ingredient in the evaluation of simulation models. The IPCC review process provides for a significant second review mechanism

---

* In November 2011, it was decided to amend the procedures and extend open invitations for expert reviewes. No separate nomination process is involved anymore, except that those who have been nominated for other tasks in the production of assessment reports are also invited as expert reviewers for chapters that they are not already involved with.

and helps lead authors to arrive at an even better grasp of the limitations of climate-simulation models. The assessment of uncertainties, for example, as carried out by the IPCC, will necessarily be a cooperative enterprise—both among individual lead authors and among lead authors and reviewers. Still, it is hard for the IPCC to do much 'better' than the scientific community. Given that the reflexivity in climate science on model structure uncertainty is relatively low (see Chapter 6), there is obviously room for improvement of IPCC's communication of uncertainty (see also Section 7.2.4).

7.2.2 Sceptical Criticism of the Review Process

Several prominent climate sceptics have challenged the integrity of the IPCC review process. They typically give examples of where they think the process has gone wrong. Here I analyse the criticism of meteorologist Richard Lindzen from the Massachusetts Institute of Technology (MIT) on the production of the TAR SPM, as a strong exemplar. In 1998, Lindzen was made part of the IPCC process as an IPCC TAR WG I lead author (of the chapter 'Physical Climate Processes and Feedbacks'; this chapter had one coordinating lead author, Thomas Stocker, and 10 lead authors). Although Lindzen was generally satisfied with the way the full report was produced, he strongly criticised the production and review process of the SPM. He testified before the U.S. Senate Commerce Committee on 1 May 2001 that many questions relevant to climate change cannot yet be answered by scientists, and that the SPM of the WG I report is not an adequate reflection of the full report. Lindzen sees himself as playing a functional role as a greenhouse sceptic. In an interview, he admitted that while in the early years of the IPCC he felt it was a 'moral obligation' to voice his sceptical opinions, 'now it is more a matter of being stuck with a role' (Grossman 2001: 36).

Most scientists would agree with Lindzen that the claims made by the IPCC are not the definitive say on the issue of climate change. This is why every few years a new assessment is produced and why IPCC WG I in the TAR has introduced in its vocabulary a gradual scale for judgmental likelihood statements, which made it possible to include an assessment of the quality of climate models in the conclusions derived from these models—albeit in a way that confuses objective and subjective probabilities, as I have shown in Chapter 6. Trying to capture controversies on the quality of models in 'consensus' judgements is tricky, of course, since the expert who thinks that the models are certainly wrong would not agree on a statement that 'there is a 10–33% chance that the models are wrong'—even if such a statement is intended to take his minority viewpoint explicitly into account. The central target of Lindzen's criticism is the published version of the detection and attribution conclusion in the WG I SPM, which was discussed in Chapters 5 and 6:

> In the light of new evidence and taking into account the remaining uncertainties, most of the observed warming over the last 50 years is

likely[7] to have been due to the increase in greenhouse gas concentrations. (IPCC 2001, SPM: 10)

Here 'likely', according to its corresponding Footnote 7, is to be read as a
66–90% chance (defined in the footnote as a 'judgmental estimate of confidence') that the statement is true. Lindzen is quite sure about the fact that the
models are wrong, and he does not trust the lead authors' judgement.

Since Lindzen has been very closely involved with the IPCC, his criticisms
merit a more detailed investigation, especially since his criticisms are related
to the assessment of climate-simulation uncertainty. In his testimony before
the Senate, Lindzen made three claims about the SPM:

1. The SPM distorts the underlying science (which is adequately represented by the chapters).
2. The SPM is 'written by representatives from governments, NGOs
   (nongovernmental organizations) and business'.
3. The SPM was significantly modified at the plenary session in
   Shanghai.

The first claim is related to the 'misrepresentation', according to Lindzen,
of computer-model uncertainty in the SPM. An example is the qualification
contained in the attribution statement just quoted (a claim that is likely true,
according to the SPM). Lindzen had not been involved in writing the SPM,
and he pointed out in his testimony that only a fraction of the lead authors
had been members of the core writing team. He did not add, however, that
the full writing team consisted of about 60 lead authors (i.e., about half of the
lead authors were involved in the drafting). The SPM representation of his
own chapter was taken by Lindzen to demonstrate his case. He claimed that
the whole chapter was summarised, inadequately, by the following sentence:

> Understanding of climate processes and their incorporation in climate
> models have improved, including water vapour, sea-ice dynamics, and
> ocean heat transport. (IPCC 2001, SPM: 9)

Lindzen's problem with this conclusion cannot be that it does not come
from the chapter since these 'improvements' were indeed all mentioned in
the chapter's Executive Summary. Furthermore, some caveats related to this
statement were put in an introductory SPM sentence immediately above the
quoted conclusion:

> [Complex physically based climate] models cannot yet simulate all
> aspects of climate (e.g., they still cannot account fully for the observed
> trend in the surface-troposphere temperature difference since 1979) and
> there are particular uncertainties associated with clouds and their inter-
> action with radiation and aerosols. (IPCC 2001, SPM: 9)

Thus, Lindzen's claim that there is only one sentence dedicated to his chapter in the SPM is not true. My contention is that Lindzen was not satisfied with the phrase 'there are particular uncertainties associated with clouds and their interaction with radiation and aerosols'. He might have preferred that the following statement was transferred from the Executive Summary of his chapter to the SPM:

> The physical basis of the cloud parameterisations included into the models has also been greatly improved. However, this increased physical veracity has not reduced the uncertainty attached to cloud feedbacks: even the sign of this feedback remains unknown. (IPCC 2001, Ch. 7: 419)

Apparently, in the face of space constraints, the lead authors who drafted the SPM had decided not to include these statements in the SPM. Here again, the issue of balance surfaces. The IPCC could have decided to include these statements and leave other statements out. It is debatable whether the fact that this did not happen must be regarded as a serious misrepresentation of science (that is, a more serious misrepresentation than any summary inevitably is).

Lindzen's second claim, that the SPM is written by nonscientific outsiders, is not true, in the sense that governments can make proposals for textual changes, but the lead authors have to agree with those changes. Indeed, the SPM drafting team (consisting only of participating scientists) paid serious attention to all comments received from experts (including experts from NGOs and business lobbies) and governments, as monitored by the review editors. As shown previously in this chapter, there were five criteria guiding the SPM writing process: scientific accuracy, scientific balance, clarity of message, understandability to policy makers, and relevance to policy. During the plenary session in Shanghai, only the different government delegations (but not the observers, as noted previously) could make comments on the text. Depending on lead authors' responses, texts were changed or left unchanged. Usually, the interventions were of such a nature that the lead authors did not have a problem with the suggested changes, that is, they agreed specifically that the suggestions were not at odds with the criteria of scientific accuracy and scientific balance, and the changes were deemed consistent with the chapters. Sometimes, the plenary was not able to reach consensus, either because the lead authors did not agree with a suggestion or because governments disagreed among themselves. Since the IPCC has to work under the procedural rule of decision making by consensus, the task of coming up with a text that was agreeable to all (including the lead authors) could in those cases be delegated to a contact group, chaired by one or more countries or by a member of the working group bureau. One of the pieces of text that was given to a contact group was the concluding statement on attribution criticised by Lindzen. This relates to Lindzen's final claim.

Lindzen's third claim is that the SPM draft was significantly modified in Shanghai. Although he did not explicitly say so, he apparently thought that the quality of the text had deteriorated because of the modifications. However, in his testimony Lindzen had made an erroneous comparison. He compared the Second Draft (April 2000) instead of the Final Draft (October 2000) to the published version, which made the change look larger than it actually was. To evaluate Lindzen's claim, I here list the four latest versions of the paragraph:

SECOND DRAFT (APRIL 2000)

From the body of evidence since IPCC *(1996), we conclude that there has been a* **discernible** *human influence on global climate.* Studies are beginning to separate the contributions to observed climate change attributable to individual external influences, both anthropogenic and natural. This work suggests that anthropogenic greenhouse gases are a substantial contributor to the observed warming, especially over the past years. However, the accuracy of these estimates continues to be limited by uncertainties in estimates of internal variability, natural and anthropogenic forcing, and the climate response to external forcing. (emphasis added in bold)

FINAL DRAFT (OCTOBER 2000)

It is likely that increasing concentrations of anthropogenic greenhouse gases have contributed **substantially** to the observed warming over the last 50 years. Nevertheless, the accuracy of estimates of the magnitude of anthropogenic warming, and particularly of the influence of the individual external factors, continues to be limited by uncertainties in estimates of internal variability, natural and anthropogenic radiative factors, in particular the forcing by anthropogenic aerosols, and the climate response to those factors. (emphasis added in bold)

SHANGHAI DRAFT (JANUARY 2001)

The precision of estimates of the contribution from individual factors to recent climate change continues to be limited by uncertainties in internal variability, natural and anthropogenic forcing, in particular that by anthropogenic aerosols, and the estimated climate response. *Despite these uncertainties, it is likely that increasing concentrations of anthropogenic greenhouse gases have contributed* **substantially** *to the observed warming over the last 50 years.* (emphasis added in bold)

APPROVED VERSION (JANUARY 2001)

In the light of new evidence and taking into account the remaining uncertainties, **most** of the observed warming over the last 50 years is likely[7] to have been due to the increase in greenhouse gas concentrations. (emphasis added in bold)

So, what has actually happened to this paragraph? The main changes in the step from Second Draft to Final Draft were the introduction of the word *likely* (incorporating both a statistical estimate of internal climate variability and an assessment of climate-model uncertainty) and the deletion of the first two sentences (they actually appeared elsewhere in the same section). The third sentence of the Second Draft became the first sentence of the Final Draft (in a more precise formulation). The following draft, the Shanghai Draft, is similar to the Final Draft except for the order of the two sentences. Finally, two changes were made during the Shanghai meeting: *Substantial* was changed to *most* and the specification of the four sources of uncertainty was removed. The phrase *remaining uncertainties* now refers to what is stated in the introductory text of the section, namely, that 'many of the sources of uncertainty identified in the SAR still remain to some degree' (IPCC 2001, SPM: 10). What happened during the Shanghai meeting was that several governments were opposed to the word *substantially*, which was therefore later replaced by *most* in a contact group meeting (for a detailed account of this meeting, see the Appendix).

It must be clear by now that I do not agree with Lindzen's negative evaluation of the review process for the SPM. Still, the detection and attribution section of the Final Draft version of the SPM was substantially changed before the Shanghai meeting, and some significant changes were *not* made in response to government comments. An example of a sentence that was not in the SPM of the Final Draft and not even in the Executive Summary of Chapter 12 is the following:

> Most of these studies find that, over the last 50 years, the estimated rate and magnitude of warming due to increasing concentrations of greenhouse gases alone are comparable with, or larger than, the observed warming. (IPCC 2001a, SPM: 10)[*]

This sentence constituted the basis for one of the most important conclusions of the IPCC (2001) report that 'there is new and stronger evidence that most of the warming observed over the last 50 years is attributable to human activities' (IPCC 2001, SPM: 10). To be sure, these statements were backed by a sentence in the chapter text itself (and by the underlying science), but—that is the point that must be made here—they were formulated by the lead authors late in the process. As prescribed by IPCC procedure, the Executive Summary of Chapter 12 was changed at the Plenary Session in Shanghai to make it consistent again with the SPM, and this change was presented to the plenary at the end of its session.

---

[*] That the estimates of the greenhouse gas contribution to the observed warming can also be larger than the observations is due to the fact that anthropogenic emissions of sulphur dioxide that lead to the formation of sulphate particles in the atmosphere have a compensating cooling effect.

7.2.3 Negotiating the Wording of the Summary for Policymakers

Uncertainties are not objectively given. Experts typically have diverging opinions about how uncertain a given statement is. Furthermore, actors with a stake in the way uncertainties are assessed and communicated by the IPCC will try to influence the formulation of the Summary for Policymakers. The positive conclusions communicated by the IPCC are taken by the governments and experts involved to be 'robust', given the assessment of uncertainties. One of the prime examples of a robust conclusion in IPCC (2001) is that 'most of the observed warming over the last 50 years is likely to have been due to the increase in greenhouse gas concentrations'. In this example, we can see that one way to ensure robustness of a conclusion is to explicitly include a qualifier within the positive statement, based on an assessment of the uncertainties involved. Here, by adding the word *likely* (and specifying what is precisely meant) the conclusion just mentioned became robust, according to the lead authors' judgement.

As was shown in Chapter 6, the main conclusion of the TAR on the attribution of climate change to human influences only implicitly reflects the collective expert judgement on the reliability$_2$—that is, the methodological reliability—of climate-simulation results. The collective assessment processes as done within the IPCC in principle provide a unique institutionalised opportunity to try to reach consensus on the models' methodological reliability and its impact on the formulation of the main attribution statements. However, this opportunity was not fully exploited—neither in the TAR analysed here nor in the AR4—partly because the IPCC is lacking a typology of uncertainty, which (if suitably chosen) would allow one to communicate unequivocally the methodological reliability of climate simulation. Still, in the production of the TAR the issue of the methodological reliability was addressed by the lead authors, and different model approaches were compared and confronted with each other, thus bringing elements of 'expert judgement' to the assessment. Furthermore, the possibility that all models have similar flaws was seriously considered.*

When the IPCC came together in Shanghai in January 2001, the robust conclusion mentioned on 'detection and attribution' could not be quickly agreed on in the plenary meeting. There were obviously political agendas behind

---

* This seems to have been less so in the production of the AR4. One sign of this was that the Final Draft SPM of the AR4 characterised the probability in AR4 statements as the 'assessed likelihood of an outcome or a result'. Only after plenary intervention by the Netherlands, this phrase was altered to read 'assessed likelihood, *using expert judgement*, of an outcome or a results' (emphasis added).

the attempts at obstruction by one country in particular, Saudi Arabia.* The argument that was used by Saudi Arabia was that the word *substantial* could not adequately be translated into Arabic, an official U.N. language.† When subsequently a delegate of France—without a similar political agenda— claimed that the translation was also problematic for his language (another official U.N. language), the chair decided to relegate the issue to a contact group meeting. The proceedings of this contact group meeting can be found in the Appendix, interspersed with my hypothetical analysis of what people were thinking when they were acting. Political agendas clearly play a role for countries in their attempts at reformulating conclusions, but these political agendas are able to force changes in the text only by referring to scientific issues or to problems with the clarity of the language. In this case, Saudi Arabia, which did have a political agenda to downplay the issue of climate change, first used the argument of clarity in the plenary session (with the translation of *substantial* purportedly being unclear) and subsequently made an issue in the contact group of the way the lead authors' assessment of computer model uncertainty was inadequately conveyed by the word *substantial*. In my judgement, since the TAR WG I report does not clearly distinguish between statistical reliability (reliability$_1$) and methodological reliability (reliability$_2$) in its formulation of robust conclusions, it was difficult for this country to separately raise the issue of the methodological reliability of models. Still, its interventions led to a significant change in the text, as can be read in the proceedings (see Appendix).

The proceedings of the contact group meeting give an interesting glimpse into the functioning of the IPCC as a boundary organisation between science and politics. From these proceedings, we can conclude that political motives leading to the use of methodological arguments can be effective in changing the text of the SPM. Most observers present in Shanghai had failed to recognise that Saudi Arabia wished to put a *quantitative modelling* statement in the conclusion of the detection and attribution section of the Summary for Policymakers of the IPCC WG I TAR, with a clear reference to that fact that it was 'only' a modelling statement, in order to be able to downplay the conclusion. The end result was probably not what Saudi Arabia really wanted. However, the lead authors had in the end accommodated the change as genuinely reflecting the contents of the underlying chapter.

---

* A formal condition for my participation as a philosophical observer in IPCC meetings was that I would ensure anonymity of countries. The meetings of the IPCC are not open to the public. Since I gained access to the meeting as a member of the Dutch government delegation, this government did not want to become responsible for a breach of bureaucratic confidentiality rules. This shows how much the IPCC is considered a diplomatic forum by governments. I have obeyed, but only for a period of 10 years. Over the past years, I have gradually come to the conviction that IPCC plenaries should be publicly accessible (e.g., through webcasting), particularly to restore trust in the organisation. In 2011, I made public that the anonymous country B (see Appendix) was actually Saudi Arabia, not a big surprise for insiders.
† The official U.N. languages into which IPCC documents are translated are Arabic, Chinese, English, French, Russian, and Spanish.

It was the difficulty of assessing and communicating computer model reliability$_2$, as compared with reliability$_1$, that caused the lead authors to pause when asked to use a relatively strong modelling statement from the body of the detection and attribution section in the conclusion. Since the word *likely* did not appear in this modelling statement, it even disappeared from view for a moment. The discussion in the contact group—more broadly, the quality of the IPCC TAR report—could have been facilitated by explicitly referring to the distinction between statistical uncertainty and methodological (un)reliability as two different sorts of uncertainty. Of course, this would not have directly solved Saudi Arabia's 'problem' with the use of computer simulation in climate science. There could still have been discussion about the methodological reliability of models and the appropriate way to communicate this reliability. But, at least the discussion would have focused on the appropriate sort of uncertainty, that is, on the methodological quality of models *as such* instead of lumping together two sorts of uncertainty.

7.2.4 IPCC Guidance Materials for Uncertainty Assessment and Communication

The problem that the lead authors of the detection and attribution statements in the IPCC TAR ran into with respect to uncertainty communication can be understood from the way IPCC WG I had implemented the IPCC guidance on uncertainty communication. In the preparation of the TAR, a strong demand for a more systematic approach to uncertainties was identified, and the subsequent discussion led to a so-called cross-cutting 'Guidance Paper' on uncertainties (Moss and Schneider 2000). In that guidance paper, Moss and Schneider proposed that authors should use a probabilistic scale that expresses subjective confidence estimates about claims in five categories: very low confidence (0–5%), low confidence (5–33%), medium confidence (33–67%), high confidence (67–95%), and very high confidence (95–100%). As a supplement to this scale, writing teams could explain their choice of category for particular claims by making use of four qualitative uncertainty expressions: 'well established', 'established but incomplete', 'competing explanations', and 'speculative'. In WG I, however, the scale proposed by Moss and Schneider was changed into a likelihood scale that was not unequivocally defined as a subjective probability scale. For the statements on climate observations, the scale was used as a purely objective frequentist scale. For the modelling statements, for instance the attribution statement extensively discussed in this chapter, the scale represented a hybrid of objective and subjective probabilities. First, a frequentist estimate was made of the chance that most of the observed warming was attributable to human influences (resulting in the very likely category, 90–99%).* Subsequently, an

---

* The confidence and likelihood scales were somewhat different for the lowest and highest categories (0–5% and 95–100% vs. 0–1% and 99–100%, respectively).

informal subjective adjustment was performed based on judgements on the unreliability$_2$ of the models, and the likelihood category likely (66–90%) was chosen. The reason for WG I to propose its own scale was that the IPCC guidance materials for the TAR lacked advice on how to represent frequentist statistical claims.

During the preparation of the IPCC AR4 (which appeared in 2007), the situation of having two separate probability scales within the IPCC was judged to be confusing, and additional guidance was prepared.* The problem that I have identified in this chapter had not been solved, however. In fact, it had become worse since lead authors were now encouraged to use only one of two scales (confidence or likelihood), without having the option to use the qualitative terminology as a supplement to these scales, as was originally proposed. The qualitative terminology was indeed only used as a substitute in AR4, in case no probabilistic statements could be made.

In the preparation of AR5, this issue surfaced again and was potentially solved in the new *Guidance Note for Lead Authors of the* IPCC *Fifth Assessment Report on Consistent Treatment of Uncertainties* (Mastrandrea et al. 2010), which was endorsed by the IPCC governments at its 33rd session (10–13 May 2011, Abu Dhabi). The *Guidance Note* prescribes that for all findings a qualitative two-dimensional level-of-understanding scale (with evidence and agreement) must be used, that no probalistic meaning must be attached to five levels of 'confidence' (very low, low, medium, high, very high), but that these levels of confidence can be used in correlation with the level of evidence and the degree of agreement. Hereby thus a consistent set of qualitative terms for the methodological reliability of findings becomes available. If well done, this could be a useful addition to the use of the likelihood scale, which like in TAR and AR4 still is designed to conflate objective and subjective probabilities. Therefore, the old rule—which was never implemented—to provide a 'traceable account' of how the author teams arrive at their uncertainty expressions should now indeed be enforced; in that way, the problem identified in this book that uncertainty characterisation has not been sufficiently transparent (cf. Meyer and Petersen 2010) could be 'solved'.†

---

* Notes for Lead Authors of the IPCC Fourth Assessment Report on Addressing Uncertainties, July 2005 (IPCC 2005).

† It is high time that the use of the qualitative level-of-understanding scale and the provision of traceable accounts becomes obligatory. It was already advised by Moss and Schneider (2000). In the first edition of the present book in 2006, I made a strong plea for it. Another reference is Swart et al. (2009), who urge for the provision of information on the 'pedigree' of findings. Finally, the IAC (2010) had advised the IPCC that all working groups should use the qualitative level-of-understanding scale in their Summary for Policymakers and Technical Summary, and that this scale may be supplemented by a quantitative probability scale, if appropriate.

7.3 An Example of Exploiting Societal Perspectives to Communicate Climate-Simulation Uncertainty

The IPCC guidance notes for dealing with uncertainty in the AR4 recommended that IPCC authors 'use neutral language' and 'avoid value laden statements' (IPCC 2005: 3). But given the unstructured nature of the problem of anthropogenic climate change, an alternative strategy to the communication of uncertainties is to consider uncertainties explicitly from different value-laden perspectives on risk in an integrated assessment of climate change. The presence in society of different perspectives on climate-change risks can be exploited in communicating the meaning of climate-simulation uncertainty to policymakers. A pioneering example of how this can be done is the TARGETS model project (Rotmans and de Vries 1997) performed by National Institute for Public Health (RIVM), precursor to the Netherlands Environmental Assessment Agency, in the mid-1990s. Note that such an integrated assessment methodology is complementary to the IPCC assessment of climate change. The IPCC assessment of climate-simulation uncertainty can be used as input for the integrated assessment.

In the TARGETS model (the acronym TARGETS stands for Tool to Assess Regional and Global Environmental and Health Targets for Sustainability), different perspectives on a large number of uncertainties about issues of sustainability were explicitly included in simulation models. The TARGETS model is an integrated assessment model that simulates both natural and social processes—and their interactions—that play a role in problems of sustainability, such as anthropogenic climate change. The timescale of interest is about a century. By building integrated assessment models, modellers can provide insight into the influence of important model uncertainties on outcomes of interest for a certain policy problem. This can be done in several ways. In the TARGETS model project, the RIVM (Rijksinstituut voor Volksgezondheid en Milieu) decided to include different plausible model structures or parameters for the *same* processes (e.g., the response of the climate system to increased CO_2 concentrations), creating different model options within the TARGETS model. Each time the TARGETS model is run, a different 'model route' can be realised, depending on pregiven specifications. In the specification of the model routes, the variation of model relations and model parameters was coupled to perspectives on risk. The way in which the TARGETS researchers did this coupling is explained in the following discussion (for more information, see van Asselt and Rotmans 1997).

The point of departure for determining the perspectives used in the TARGETS model was a simplified version of 'cultural theory'. Cultural theory has been developed by Mary Douglas, Michael Thompson, and Aaron Wildavasky, among others (see, e.g., Thompson et al. 1990). In Douglas's original 'grid-group theory' (the precursor of cultural theory), it was assumed that the variability in the way individuals take part in social life can adequately be described based on two dimensions (those of grid and group). The *group* dimension refers to the presence of social ties with a certain social group. The *grid* dimension refers to the freedom of individuals to make choices (belonging to a group does not necessarily limit individual freedom of choice). The assumption is made in grid-group theory that these dimensions are independent from each other. Thompson et al. (1990: 26–29) couple the way individuals take part in social life to different 'myths of nature'.* In the TARGETS model, three different perspectives are distinguished: the 'individualist' (low group, high grid); the 'egalitarian' (high group, high grid); and the 'hierarchist' (high group, low grid) perspectives. The two remaining perspectives distinguished within cultural theory, namely, the 'fatalist' (low group, low grid) and the 'hermit' (middle group, middle grid) perspectives, are left aside in the TARGETS model. The perspectives as they have been used in the TARGETS model concern, on the one hand, visions on nature and humans (how nature and society work—called 'world views'), and, on the other hand, preferred management styles (how we should govern). It must be remembered that the perspectives used are caricatures. Sometimes, these perspectives are recognisable in debates, but individual people cannot be strictly categorised in terms of the notions of cultural theory.

In the *individualist* perspective, nature is robust against human perturbations. In this perspective, people are oriented toward (rationally) satisfying their own wants, and they think 'anthropocentrically'. The individualist management style consists of adapting policy to changes in natural and societal conditions, aiming toward strong economic growth and knowingly taking risks (thus, on the one hand, the perception of risks is coloured by world views and, on the other hand, dealing with these perceived risks is coloured by the preferred management style). Since the TARGETS model operates at a high level of aggregation (the variables have been defined at the regional or global level), the individualist management style has been implemented as the 'top-down' governance by a world governor (who wishes to facilitate that people can hold their individualist perspective) instead of it being modelled as the resultant of parties that operate as individualists ('bottom-up' approach).

As opposed to the individualist perspective, from the *egalitarian* perspective nature is vulnerable. Small perturbations can lead to catastrophes. Human influences on ecosystems are considered to be significant disruptions and are dangerous in most cases. In this perspective, humans are good by nature, but they are malleable. Relationships that are based on equality

---

* Thompson et al. (1990: 28) state that they 'have deduced [these myths] by asking [respondents] how nature would have to be conceived for our ways of life to be livable'. It is not well established that indeed people's visions on nature and visions on human beings and society are really coupled, as is assumed in this version of cultural theory. This is still an open question for sociologists working on this topic.

with nature and fellow human beings evoke the goodness of people. The egalitarian perspective is connected with 'ecocentric' thought. This also becomes evident from the egalitarian management style, which is aimed at precaution and can be characterised as risk avoiding.

Finally, a *hierarchist* perspective is characterised by a conception of nature as robust within particular limits. In the hierarchist perspective, individual humans are 'sinful' and can be redeemed by means of the appropriate societal institutions (aimed at both human well-being and nature quality). The hierarchist perspective is connected with 'participating' thought, an attitude of partnership with nature (thus neither anthropocentrism nor ecocentrism). In the hierarchist management style, controlled economic growth is allowed, and the taking of particular risks is accepted. The limits of the stability of nature are systematically studied, and one tries to stay within the limits found.

In defining the different model routes that can be followed in the TARGETS model, the perspectives' assumptions of how nature and society work have been translated into scientific assumptions about model structures and model parameters.* For the model parameter 'climate sensitivity' (described in Chapter 5), for example, the individualist (think of the coal and oil lobby) chooses a value in the lower end of the range of uncertainty (which is a 1.5°C equilibrium temperature increase for a doubling of the CO_2 concentration), and the egalitarian (think of the environmental lobby) chooses a value in the upper end of this range (being 4.5°C). The hierarchist, finally, chooses a value that lies between these two values. Note that the range of 1.5–4.5°C is the current IPCC uncertainty range for climate sensitivity.

Subsequently, the model can be run for this century using nine different combinations of world views and management styles, producing a set of future scenarios, in three of which the world behaves in the way the management style assumes it to behave and in six of which there is a mismatch between real-world behaviour and management style. We can, for instance, see what happens if the egalitarian world view is correct but an individualist management style is chosen for governing the world. From running such a scenario, we can learn that all limits that should be met—if the egalitarian worldview is indeed true—will be excessively exceeded.

The TARGETS model approach thus delivers more than just an estimate of model uncertainty. It makes it possible to attach meaning to the model results for the nine scenarios that result from this approach. These nine

---

* The way the perspectives are connected in the TARGETS model to alternative scientific assumptions can give the impression that there is a simple relation between a scientist's political view and his or her scientific beliefs. Indeed, van Asselt and Rotmans (1997) refer to the sociology of scientific knowledge to argue for the plausibility of such a connection. However, it is difficult to find evidence for this position in scientific practice (see, e.g., Pickering 1992). Therefore, I propose to interpret the coupling as relationships between societal perspectives on risk and what these entail in terms of scientific assumptions (within a range of plausible and established scientific assumptions) rather than as a claim that political views would determine the development of scientific knowledge.

scenarios can be divided into two groups. The first group of scenarios are three 'utopias', in which world views and management styles belong to the same perspective.* These scenarios show the future developments (of, for instance, global temperature change) associated with three possible worlds, in which the management styles chosen fit the real workings of nature and human behaviour. The range of these three model results can be interpreted as a first estimate of the simulation uncertainty of the integrated assessment model (this uncertainty obviously encompasses many more uncertainties than those related to the simulation of climate). The second group of scenarios is six 'dystopias'. In the dystopias, there is a mismatch between the factual behaviour of nature and humans, on the one hand, and the management style chosen on the other hand. The outcomes of the dystopias can help policymakers to judge the three management styles.

How does this relate to the issue of climate change and climate-simulation uncertainty? The TARGETS researchers have produced nine scenarios for climate change (see den Elzen et al. 1997). When the egalitarian worldview is combined with either the individualist or the hierarchist management style, one finds that the upper limit for the CO_2 concentration according to the egalitarian world view (450 ppmv, or particles per million by volume, meaning 450 CO_2 molecules in 1 million air molecules; the preindustrial concentration was 280 ppmv) will be seriously exceeded from 2050 onwards. Thus from the egalitarian perspective, both other management styles entail large risks. In contrast, if the individualist world view is true (assuming that 650 ppmv is safe), none of the management styles leads to exceeding this norm before the end of the century.

The advantage of using integrated assessment models in science-for-policy is that models from different scientific disciplines (in the case of TARGETS, models of the physical climate system, energy, population, among others) are combined. The intention of such models is to produce useful information for policymakers—information that is expected to have added value as compared to monodisciplinary knowledge. Apart from the advantages of this kind of model, the disadvantages of most integrated assessment models—as assessed by Rotmans and Dowlatabadi (1998)—are that (1) they are very complex; (2) they use a high level of aggregation; (3) they lack credibility in the different scientific disciplines; (4) they treat uncertainties in an inadequate manner (the TARGETS model being an exception); (5) they apply a deterministic paradigm; (6) they have only been marginally verified and validated (as far as this is possible at all); (6) they contain inadequate knowledge; and (7) they are limited in the modelling formalism by the application of only a few standard methods. The TARGETS model has been able to remove some of these disadvantages (notably 1, 3, 4). The advantages offered

---

* The terms *utopia* and *dystopia* (see the following discussion) are used by the TARGETS modellers as technical terms to denote a match or mismatch, respectively, between the assumed and real behaviour of the world.

by integrated assessment models are that they allow—even if only at a rudimentary level—the study of interactions and feedbacks that cannot be studied otherwise; that they are flexible and fast; and that they can help intensify the communication between scientists and policymakers. To conclude, even though some methodological challenges remain, the TARGETS approach is definitely attractive for making connections between scientific information and political choices. It can assist policymakers in making choices under climate-simulation uncertainty. However, it is important that such models are not used as substitutes for political decision making: They should be regarded merely as quantitative tools to communicate uncertainties.*

7.4 Conclusion

In this chapter, I analysed two examples of the communication of climate-simulation uncertainty in policy advice. These are examples of assessments of climate change that were performed at different scales: one by an intergovernmental organisation (IPCC) and another by a national agency (the precursor of the Netherlands Environmental Assessment Agency).

The IPCC assessments are social constructs that contain both scientific and political elements. The IPCC's success depends on its ability to connect to both climate science and climate policy. The generally voiced criticism that the IPCC is not open enough to 'sceptics' is largely untrue. The IPCC procedures ensure inclusivity, and sceptics do have influence on the formulation of the reports. However, the IPCC could be much more reflexive on its procedures and ways to deal with dissensus. All in all, I would not characterise the IPCC reports as constituting the 'scientific consensus' on climate change, but instead as 'policy-relevant assessments acknowledging uncertainty'.

The IPCC uses value-neutral statistical expressions of uncertainty (e.g., it is 'likely' that most of the observed warming is due to human influences, or, the climate sensitivity is between 1.5°C and 4.5°C), while the TARGETS approach uses value-laden scenario expressions of uncertainty (e.g., an egalitarian assumption on climate sensitivity is 4.5°C, while an individualist assumption is 1.5°C). As approaches for scientists to offer policy advice on unstructured policy problems, they have different characteristics. The IPCC strategy, notably that of WG I, tends to downplay value-ladenness and is therefore able to maintain a clear boundary between science and policy while involving policymakers in its deliberations on the wording of summaries for

---

* Compared with an earlier publication (Petersen 2000b), I have become more positive about the merits of perspective-based approaches. The Netherlands Environmental Assessment Agency has applied a somewhat similar approach in its Sustainability Outlook of 2004 (see de Vries and Petersen 2009).

policymakers. TARGETS offers the possibility to explicitly discuss values in relation to climate risks (what bets are we willing to take on the magnitude of climate sensitivity?).

Neither of these strategies is ideal, but they both need to be pursued since they are both able to communicate climate-simulation uncertainty to policy-makers, albeit in different manners. Meanwhile, improvements need to be made to both strategies. The IPCC will have to become more transparent in its assessment of uncertainty (e.g., Why is a particular likelihood category chosen? What is the unreliability$_2$ of the underlying climate simulations?) and will have an opportunity to do so in its Fifth Assessment Report, due in 2013–2014, and onwards. And, the Netherlands Environmental Assessment Agency will have to replace its too schematic perspectives (worldviews) based on cultural theory by perspectives derived from the interaction with policymakers and stakeholders to tailor its perspective-based assessment and modelling activities to the audiences that play a role in the political and societal debate.

8

Conclusions

8.1 Uncertainty Associated with Scientific Simulation

The basic tenets of this study are that there is a plurality of simulation models, methodologies, and values in simulation practice; that the kind and reasonableness of this plurality can be understood philosophically; and that this plurality should be cherished. This plurality is important for scientific reasons and, where simulation is used in policymaking, for sociopolitical reasons. In conclusion, I take stock of the results obtained in the various chapters and determine what answers can be given to the five research questions posed in Chapter 1.

The first research question was: What specific types of uncertainty are associated with scientific simulation? By following a strategy similar to Giora Hon's (1989, 2003), who arrived at the structural features of experiment through a philosophical analysis of the notion of error in experimentation, I have proposed four central elements of simulation that correspond to the locations where uncertainties arise in simulation practice. In this way, it has been possible to 'transcend the "etc. list"' (Hon 2003) of the myriad elements (strategies, methods, procedures, conceptions, styles, etc.) that constitute this practice. The central elements of simulation are (1) the conceptual and mathematical model; (2) the model inputs; (3) the technical model implementation; and (4) the processed output data and their interpretation. By analysing these elements, I have been able to clarify what is specific about the types of uncertainty that are associated with scientific simulation, as compared with other scientific practices.

In addition, to arrive at a complete characterisation of the uncertainties involved in simulation, I have extended Funtowicz and Ravetz's (1990) typology of uncertainty and presented five uncertainty dimensions besides the dimension of location: (1) nature of uncertainty; (2) range of uncertainty; (3) recognised ignorance; (4) methodological unreliability; and (5) value diversity.

In the discussion of the four main elements of simulation practice, four philosophical issues with respect to simulation were addressed, which enable some general characterisations of simulation uncertainty to be formulated.

1. I argued against Cartwright (1983) that the distinction between general theory and models should be considered to be a relative one: Some theoretical equations that are considered as a model from the perspective of a more fundamental (sub)discipline can also be considered as a general theory from which approximate models are derived. Simulation models are not fully derived from theory, however. For instance, most simulation models of complex systems contain a number of 'parameterisations' of processes that cannot be simulated in more detail. These parameterisations are typically not fully based on general theory. By determining both the extent to which simulation models are derived from general theory and the scope of the general theory, one can assess the theoretical quality of simulations.

2. I emphasised that the accuracy of a simulation can be increased by using real-world input (following Morgan 2003), but that the extent to which the outcomes are reliable depends not only on the input data, but also on the reliability of the conceptual and mathematical model (an issue that is emphasised too little by Morgan).

3. Extending Radder's (1996) account of experimental reproducibility, I argued that reproducing simulation runs by using one technical model implementation on the same computer system is typically unproblematic, while transferring computer models to other computer systems, or building new model implementations, is more difficult. The last strategies can provide checks on the effects of numerical approximations and mistakes in the original computer programs.

4. In Section 2.3.4, I showed that by using advanced visualisation techniques, including animation, simulationists can come to better understand the processes under study, provided that they keep track of the influence of the projection methods chosen to produce the pictures. These projection methods are to a large extent arbitrary, leading to the risk of incorrect conclusions.

Some conclusions on simulation uncertainty can also be drawn from the discussion of each of the five uncertainty dimensions besides the location dimension.

1. Although the nature of simulation uncertainty can be ontic, there is usually an epistemic uncertainty about this ontic uncertainty. Epistemic uncertainty is a consequence of the incompleteness and fallibility of knowledge (as included, for instance, in the model structure or the model inputs). In principle, simulation models can simulate ontic uncertainty, that is, the intrinsically indeterminate or variable character of the system under study. The uncertainty of

assumptions in the model, however, results in epistemic uncertainty about ontic uncertainty.

2. Uncertainty ranges of two types can be determined for the results of simulations: statistical uncertainty ranges and scenario uncertainty ranges. Statistical uncertainty ranges can be determined either from comparing the simulation results with measurements (provided that accurate and sufficient measurements are available) or from uncertainty analysis (provided that the accuracy of the different elements in simulation is known). Scenario uncertainty ranges (based on 'what-if' questions) are generally more easily constructed by varying simulation elements.

3. Since simulation models are simplified representations of reality, simulationists who study complex systems must be aware of the fact that important processes that may strongly influence the nonlinear behaviour (e.g., nonlinear feedbacks) of these systems may be poorly represented (or even absent) in the models. More generally, we realise—in one way or another—that some uncertainties are present, but we cannot establish any useful estimate, such as due to limits of predictability and knowability ('chaos') or due to unknown processes. This ignorance should be recognised, and in our claims based on simulation results, we must be open about it.

4. It is often not possible to establish the accuracy of the results of a simulation. The methodological rigour of the simulation may then be regarded as an alternative for accuracy as an estimate of the reliability (then denoted by reliability$_2$) of a simulation. Four criteria distinguished for this purpose are theoretical basis, empirical basis, comparison with other simulations, and acceptance/support within and outside the peer community. First, it is important to determine both the extent to which simulation models are derived from general theory and the scope of the general theory. Second, it should be assessed to what extent simulations are based on or have been tested against observations or experiments. The qualitative assessment of the fit between the simulation results and the system of study also belongs to the assessment of the empirical basis. Third, the results of simulations should be related to the results of other simulations of similar processes to determine whether the results are replicable. Fourth, the simulation should be peer reviewed, at least internally within a discipline and externally if this is relevant.

5. Since simulationists often have considerable freedom in making choices, simulations inevitably have a subjective component and may be influenced by epistemic and nonepistemic values held by the simulationist. The general epistemic, disciplinary-bound epistemic, sociopolitical, and practical values shaping a simulation need to be assessed.

The uncertainty typology that I used in this study incorporates a broad notion of uncertainty. It is, in particular, much broader than statistics. By highlighting the dimensions of methodological unreliability and value diversity, I have been able to give a more adequate account of simulation practice and its uncertainties than would have been possible by extending error-statistical accounts of experimental practice (such as Mayo 1996) to simulation practice.

8.2 Differences and Similarities between Simulation and Experimental Uncertainty

The second research question was: What are the differences and similarities between simulation uncertainty and experimental uncertainty? In Sections 2.6 and 3.8, the practices and uncertainties of simulation and experimentation were compared, and differences and similarities were identified. My claim is that, although there are similarities between simulation and experimentation, the differences are fundamental. I have dealt with four questions that can help in answering the second research question:

1. Is simulation a form of experimentation?
2. How close can simulation come to material experiments?
3. Are both simulation and experimentation forms of modelling?
4. How does the norm of reproducibility in simulation practice compare with the same norm in experimental practice?

I here answer these questions based on the results gained in the present study.

Is simulation a form of experimentation? Simulations are often characterised by philosophers and social students of science as 'experiments' on theories (e.g., Rohrlich 1991; Humphreys 1994; Galison 1996; Dowling 1999; Keller 2003; Morgan 2003). By analysing simulation as a laboratory practice, the skills involved in 'experimenting', 'playing around', 'tinkering', and so on with simulation models show some superficial analogies to the skills in experimental practice. However, the elements of simulation practice are different from those in experimental practice. The main difference is that what is manipulated in simulation practice are mathematical models materialised in computer programs and not material models. Thus the skills are of a different nature, and the uncertainties corresponding to material elements are of a different nature. I consider the difference between mathematical and material manipulations to be fundamental, and I therefore propose not to associate the term *experiment* with simulation.

How close can simulation come to material experiments? Morgan (2003) claims that some simulations look more like material experiments than others (based on the verisimilitude of the input data to reality) and hence are less uncertain than other simulations. However, as I have shown, this proximity between simulation and experimentation should only be interpreted as closeness of the mathematical input of the simulation to the material input of the material experiment. It does not entail closeness of the mathematical model and the model results to reality.

Are both simulation and experimentation forms of modelling? Both simulation and experimental practices involve models of the outside world and extrapolation of the results obtained in the laboratory to this outside world. Thus both practices are confronted with the different types of uncertainties in models. This similarity between simulation and experimentation is important since when models in experiments remain hidden from view, the fact that the results of measurements are sensitive to modelling assumptions can be obscured. However, the representational relationship between a mathematical model and reality is different in kind from the representational relationship between a material model and reality. Experiments are confronted with more types of uncertainty than simulations.

How does the norm of reproducibility in simulation practice compare with the same norm in experimental practice? The norm of the reproducibility pertains to both simulation and experimental practice. However, the activities involved in meeting this norm are not the same, and generally the norm is more difficult to meet in experimental practice than in simulation practice. Again, experimental practice involves more types of uncertainty that have to be overcome to obtain reproducibility.

From the main difference between simulation and experimentation, namely, that the physical nature that is under study is not present in the simulation laboratory, we can conclude that there are additional types of uncertainty in experimental practice. In material experiments, we may be 'confounded' by a behaviour of the natural processes that is different from what we expected, while we can only be 'surprised' by the properties of mathematical models. But although there may be fewer types of uncertainty in simulation practice, the model uncertainty in simulation may well be larger than all uncertainties involved in experimentation. Since this need not be so, however, I conclude that the question of whether uncertainties in experimentation are smaller than uncertainties in simulation has to be decided on a case-by-case basis.

8.3 Assessment and Communication of Scientific Simulation Uncertainties in Science-for-Policy

The third research question was: What are appropriate ways to assess and communicate scientific simulation uncertainties in science-for-policy? In Chapter 4, I argued that since policymakers are usually not themselves able to judge the uncertainty of scientific simulation-model outcomes, scientific policy advisers must carefully weigh how to present their conclusions. I agreed with Funtowicz and Ravetz that policy problems with high societal stakes and high scientific uncertainty attached to them require 'postnormal science' as a problem-solving strategy, and I emphasised that postnormal science poses a special challenge to scientific advisers to assess and communicate uncertainties appropriately and responsibly.

I showed that the way uncertainties in simulation should be dealt with by scientific advisers depends on the type of policy problem that confronts them. Four types of policy problems were therefore distinguished based on literature in the area of political science (Ezrahi 1980; Hisschemöller and Hoppe 1996; Hisschemöller et al. 2001). These policy-problem types vary in the level of agreement on the political objectives and on the scientific knowledge relevant to the problem. I claimed that the *Guidance on Uncertainty Assessment and Communication* (Petersen et al. 2003; Janssen et al. 2003; van der Sluijs et al. 2003, 2004) introduced by the Netherlands Environmental Assessment Agency, which incorporates these insights from political science, can be used as a practical methodology to determine what information on simulation uncertainty is appropriate to be included in policy advice.

The analysis of simulation uncertainty in science that was presented in Chapters 2 and 3 (see Section 8.1) provides relevant insights that should be taken into account when assessing and communicating scientific simulation uncertainties in science-for-policy.

8.4 Uncertainty Associated with the Simulation-Based Attribution of Climate Change to Human Influences

In the climate case, for instance, one can find different simulation models that make mutually conflicting claims about what the climate system is like.

Climate scientists are unable to identify which climate model actually incorporates the most realistic assumptions about the climatic system. What the uncertainty typology of this study offers them is a means to assess dimensions of uncertainty other than inaccuracy. In particular, the dimensions of recognised ignorance and methodological unreliability need their systematic attention.

The fourth research question was: What specific types of uncertainty are associated with the simulation-based attribution of climate change to human influences? and was studied as a historical case with the state of the science of the year 2000. In general, the parameterisation of clouds constitutes the most important source of uncertainty in climate models. For the attribution of the observed global warming since the middle of the 20th century to human influences, the five key uncertainties were the internal climate variability (an ontic uncertainty simulated by climate models, hence giving rise to epistemic uncertainty about this ontic uncertainty); the natural forcing (located in the model inputs); the anthropogenic forcing (through the emission of greenhouse gases, among other factors); the response patterns to natural and anthropogenic forcing; and the free atmosphere temperature trends. For all of these key uncertainties, it was crucial to assess and communicate the recognised ignorance and methodological unreliability. Meanwhile, scientists have found out that the last uncertainty, about free atmosphere temperature trends, does not constitute one of the most important uncertainties for the models anymore since the disagreement with the observations was resolved by improving the observations.

By assessing all these uncertainties, the Intergovernmental Panel on Climate Change (IPCC) reached the conclusion in 2001 that it is 'likely' (between 66% and 90% chance) that the largest part of the observed warming is due to human influences. The simulation uncertainty thus does not preclude policy-relevant statements about climate change. Whether the said conclusion was the result of an appropriate assessment and communication of uncertainty is discussed in the next section.

8.5 Assessment and Communication of Attribution Uncertainty by the IPCC

The fifth research question was: Have these uncertainties been appropriately assessed and communicated by the IPCC? The IPCC lacks a typology of uncertainty that can be used to assess uncertainties systematically. The typology of simulation uncertainty proposed in this study can be fruitfully applied in the analysis of climate-simulation uncertainty, as was shown for the simulation-related sources of uncertainty in climate-change attribution studies.

By applying the typology, it became immediately obvious that only part of the uncertainty can be expressed statistically. Additional qualitative judgements on the reliability$_2$ of the climate-simulation models are needed—and indeed played an important role in the production of the IPCC (2001) report, the historical case under study. Since the vocabulary needed to distinguish explicitly between the two uncertainty sorts of unreliability$_1$ (inaccuracy) and unreliability$_2$ (methodological unreliability) was not available to the lead authors, the influence of their qualitative judgements on reaching their final conclusion remained largely invisible to outsiders. Still, the review process of the Third Assessment Report and IPCC's assessment of uncertainty were evaluated positively in this study. However, the IPCC's communication of uncertainty still needs further improvement. In the Fifth Assessment Report (AR5), to be published in 2013–2014, opportunities arise for such improvement in the implementation of the new *Guidance Note* of 2010 (Mastrandrea et al. 2010).

As an alternative to communicating climate-simulation uncertainty via value-neutral statistical expressions of uncertainty, as is done by the IPCC, I sketched the TARGETS (Tool to Assess Regional and Global Environmental and Health Targets for Sustainability) approach, a perspective-based integrated assessment methodology to communicate uncertainties within a risk framework. Both the IPCC and the TARGETS strategies were found to have advantages and disadvantages. It follows from this study that both strategies can be used in a complementary manner. The IPCC procedure can guide the assessment and expression of the range of uncertainty, and via perspective-based risk assessment, this range can be made more meaningful for policymakers.

References

Agrawala, S. (1998), Context and early origins of the Intergovernmental Panel on Climate Change, *Climatic Change* 39: 605–620.

Apostel, L. (1961), Towards the formal study of models in the non-formal sciences, in H. Freudenthal (ed.), *The Concept and the Role of the Model in Mathematics and Natural and Social Sciences*, Dordrecht, the Netherlands: Reidel, pp. 1–37.

Baede, A.P.M., Ahlonsou, E., Ding, Y., and Schimel, D. (2001), The climate system: an overview, in IPCC, *Climate Change 2001: The Scientific Basis. Contribution of Working Group I to the Third Assessment Report of the Intergovernmental Panel on Climate Change*, Cambridge, England: Cambridge University Press, pp. 85–98.

Bailer-Jones, D., and Hartmann, S. (1999), Modell, in H.J. Sandkühler (ed.), *Enzyklopädie Philosophie*, Hamburg, Germany: Felix Meiner Verlag, pp. 854–859.

Balci, O. (1994), Validation, verification, and testing techniques throughout the life cycle of a simulation study, *Annals of Operations Research* 53: 121–173.

Barber, B. (1984), *Strong Democracy: Participatory Politics for a New Age*, Berkeley: University of California Press.

Benammar, K.J. (1993), Pictures of thought: The representational function of visual models, Ph.D. dissertation, University Park: Pennsylvania State University.

Boehmer-Christiansen, S. (1994a), Global climate protection policy: the limits of scientific advice, part 1, *Global Environmental Change* 4: 140–159.

Boehmer-Christiansen, S. (1994b), Global climate protection policy: the limits of scientific advice, part 2, *Global Environmental Change* 4: 185–200.

Boumans, M. (1999), Built-in justification, in M.S. Morgan and M. Morrison (eds.), *Models as Mediators: Perspectives on Natural and Social Science*, Cambridge, England: Cambridge University Press, pp. 66–96.

Bray, D., and von Storch, H. (1999), Climate science: an empirical example of postnormal science, *Bulletin of the American Meteorological Society* 80: 440–455.

Bray, D., and von Storch, H. (2007), *Climate Scientists Perceptions of Climate Change Science*, GKSS Report 2007/11, Geesthacht, Germany: Institute for Coastal Research, GKSS Forschungszentrum. Available at: http://dvsun3.gkss.de/BERICHTE/GKSS_Berichte_2007/GKSS_2007_11.pdf.

Brown, G.E. (1996, October 23), *Environmental Science under Siege: Fringe Science and the 104th Congress*, Report to the Democratic Caucus of the Committee on Science, U.S. House of Representatives. Available at: http://archives.democrats.science.house.gov/Media/File/Reports/environment_science_report_23oct96.pdf.

Brown, H.I. (1988), *Rationality*, London: Routledge.

Cartwright, N. (1983), *How the Laws of Physics Lie*, Oxford, England: Clarendon Press.

Cartwright, N. (1989), *Natures Capacities and Their Measurement*, Oxford, England: Clarendon Press.

Cartwright, N. (1999), Models and the limits of theory: quantum Hamiltonians and the BCS models of superconductivity, in M.S. Morgan and M. Morrison (eds.), *Models as Mediators: Perspectives on Natural and Social Science*, Cambridge, England: Cambridge University Press, pp. 241–281.

Casti, J.L. (1997), *Would-Be Worlds: How Simulation Is Changing the Frontiers of Science*, New York: Wiley.

Claussen, M., Mysak, L.A., Weaver, A.J., Crucifix, M., Fichefet, T., Loutre, M.-F., Weber, S.L., Alcamo, J., Alexeev, V.A., Berger, A., Calov, R., Ganopolski, A., Goosse, H., Lohmann, G., Lunkeit, F., Mokhov, I.I., Petoukhov, V., Stone, P., and Wang, Z. (2002), Earth system models of intermediate complexity: closing the gap in the spectrum of climate system models, *Climate Dynamics* 18: 579–586.

Climate Change Science Program (CCSP) (2009), *Best Practice Approaches for Characterizing, Communicating, and Incorporating Scientific Uncertainty in Climate Decision Making* (M.G. Morgan [Lead Author], H. Dowlatabadi, M. Henrion, D. Keith, R. Lempert, S. McBride, M. Small, and T. Wilbanks [Contributing Authors]), A Report by the Climate Change Science Program and the Subcommittee on Global Change Research, Washington, DC: National Oceanic and Atmospheric Administration, 96 pp.

Collingridge, D., and Reeve C. (1986), *Science Speaks to Power*, New York: St. Martin's Press.

de Kwaadsteniet, H. (1999, January 20), De samenleving heeft recht op eerlijke informatie, het RIVM geeft die niet [Society has a right to honest information, which RIVM does not provide], *Trouw*, p. 16. [In Dutch]

den Elzen, M.G.J., Beusen, A.H.W., Rotmans, J., and van Asselt M.B.A. (1997), Human disturbance of the global biogeochemical cycles, in J. Rotmans and H.J.M. de Vries (eds.), *Perspectives on Global Change: The TARGETS Approach*, Cambridge, England: Cambridge University Press, pp. 347–370.

de Regt, H.W. (1996), Are physicists philosophies irrelevant idiosyncrasies? *Philosophica* 58: 125–151.

de Regt, H.W., and Dieks, D. (2005), A contextual approach to scientific understanding, *Synthese* 144: 137–170.

de Vries, H.J.M., and Petersen, A.C. (2009), Conceptualizing sustainable development: an assessment methodology connecting values, knowledge, worldviews and scenarios, *Ecological Economics* 68: 1006–1019.

Douglas, H. (2004), The irreducible complexity of objectivity, *Synthese* 138: 453–473.

Dowling, D. (1999), Experimenting on theories, *Science in Context* 12: 261–273.

Edwards, P.N. (1999), Global climate science, uncertainty and politics: data-laden models, model-filtered data, *Science as Culture* 8: 437–472.

Edwards, P.N., and Schneider, S.H. (2001), Self-governance and peer review in science-for-policy: the case of the IPCC Second Assessment Report, in C.A. Miller and P.N. Edwards (eds.), *Changing the Atmosphere: Expert Knowledge and Environmental Governance*, Cambridge, MA: MIT Press, pp. 219–246.

Errico, R.M. (2000), On the lack of accountability in meteorological research, *Bulletin of the American Meteorological Society* 81: 1333–1337.

Ezrahi, Y. (1980), Utopian and pragmatic rationalism: the political context of scientific advice, *Minerva* 18: 111–131.

Frigg, R., and Reiss, J. (2009), The philosophy of simulation: hot new issues or same old stew?, *Synthese* 169: 593–613.

Fujimura, J.H. (1992). Crafting science: standardized packages, boundary object, and 'translation', in A. Pickering (ed.), *Science as Practice and Culture*, Chicago: Chicago University Press, pp. 168–211.

Funtowicz, S.O., and Ravetz, J.R. (1985), Three types of risk assessment: a methodological analysis, in C. Whipple and V.T. Covello (eds.), *Risk Analysis in the Private Sector*, New York: Plenum Press, pp. 217–231.

Funtowicz, S.O., and Ravetz, J.R. (1990), *Uncertainty and Quality in Science for Policy*, Dordrecht, the Netherlands: Kluwer Academic.

Funtowicz, S.O., and Ravetz J.R. (1991), A new scientific methodology for global environmental issues, in R. Constanza (ed.), *Ecological Economics: The Science and Management of Sustainability*, New York: Columbia University Press, pp. 137–152.

Funtowicz, S.O., and Ravetz, J.R. (1993), Science for the post-normal age, *Futures* 25: 739–755.

Galison, P. (1996), Computer simulations and the trading zone, in P. Galison and D.J. Stump (eds.), *The Disunity of Science: Boundaries, Contexts, and Power*, Stanford, CA: Stanford University Press, pp. 118–157.

Galison, P. (1997), *Image and Logic: A Material Culture of Microphysics*, Chicago: Chicago University Press.

Gardiner, S.M. (2004), Ethics and global climate change, *Ethics* 114: 555–600.

Gieryn, T.F. (1999), *Cultural Boundaries of Science: Credibility on the Line*, Chicago: Chicago University Press.

Goldenfeld, N., and Kadanoff, L.P. (1999), Simple lessons from complexity, *Science* 284: 87–89.

Grossman, D. (2001, November), Dissent in the maelstrom: Profile Richard S. Lindzen, *Scientific American* 285(5): 36–37.

Guston, D.H. (2001), Boundary organizations in environmental policy and science: an introduction, *Science, Technology, and Human Values* 26: 399–408.

Haag, D., and Kaupenjohann, M. (2001), Parameters, prediction, post-normal science and the precautionary principle: a roadmap for modelling for decision-making, *Ecological Modelling* 144: 45–60.

Haas, P.M. (1990), *Saving the Mediterranean: The Politics of International Environmental Cooperation*, New York: Columbia University Press.

Hacking, I. (1992), The self-vindication of the laboratory sciences, in A. Pickering (ed.), *Science as Practice and Culture*, Chicago: University of Chicago Press, pp. 29–64.

Hage, M., Leroy, P., and Petersen, A.C. (2010), Stakeholder participation in environmental knowledge production, *Futures* 42: 254–264.

Harper, K.C. (2003), Research from the boundary layer: civilian leadership, military funding and the development of numerical weather prediction (1946–55), *Social Studies of Science* 33: 667–696.

Harré, R. (2003), The materiality of instruments in a metaphysics for experiments, in H. Radder (ed.), *The Philosophy of Scientific Experimentation*, Pittsburgh, PA: Pittsburgh University Press, pp. 19–38.

Hartmann, S. (1996), The world as a process, in R. Hegselmann, U. Mueller, and K.G. Troitzsch (eds.), *Modelling and Simulation in the Social Sciences from the Philosophy of Science Point of View*, Dordrecht, the Netherlands: Kluwer Academic, pp. 72–100.

Harvey, L.D.D., Gregory, L., Hoffert, M., Jain, A., Lal, M., Leemans, R., Raper, S., Wigley, T., and de Wolde, J. (1997), *An Introduction to Simple Climate Models Used in the IPCC Second Assessment Report*, IPCC Technical Paper 2, Geneva: Intergovernmental Panel on Climate Change. Available at: http://www.ipcc.ch/pdf/technical-papers/paper-II-en.pdf.

Held, I.M. (2005), The gap between simulation and understanding in climate model-ing, *Bulletin of the American Meteorological Society* 86: 1609–1614.

Hesse, M.B. (1963), *Models and Analogies in Science*. London: Sheed and Ward.

Hisschemöller, M., and Hoppe, R. (1996), Coping with intractable controversies: the case for problem structuring in policy design and analysis, *Knowledge and Policy* 8: 40–60.

Hisschemöller, M., Hoppe, R., Groenewegen, P., and Midden, C.J.H. (2001), Knowledge use and political choice in Dutch environmental policy: a problem-structuring perspective on real life experiments in extended peer review, in M. Hisschemöller, R. Hoppe, W.N. Dunn, and J.R. Ravetz (eds.), *Knowledge, Power, and Participation in Environmental Policy Analysis*, New Brunswick, NJ: Transaction, pp. 437–470.

Hon, G. (1989), Towards a typology of experimental errors: an epistemological view, *Studies in History and Philosophy of Science* 20: 469–504.

Hon, G. (2003), The idols of experiment: transcending the 'etc. list', in H. Radder (ed.), *The Philosophy of Scientific Experimentation*, Pittsburgh, PA: Pittsburgh University Press, pp. 174–197.

Humphreys, P. (1991), Computer simulations, in A. Fine, M. Forbes, and L. Wessels (eds.), *PSA 1990, Vol. 2: Symposium and Invited Papers*, East Lansing, MI: Philosophy of Science Association, pp. 497–506.

Humphreys, P. (1994), Numerical experimentation, in P. Humphreys (ed.), *Patrick Suppes: Scientific Philosopher, Vol. 2*, Dordrecht, the Netherlands: Kluwer Academic, pp. 103–121.

Humphreys, P. (1995), Computational science and scientific method, *Minds and Machines* 5: 499–512.

InterAcademy Council (IAC) (2010), *Climate Change Assessments: Review of the Processes and Procedures of the IPCC*, Amsterdam: InterAcademy Council. Available at: http://reviewipcc.interacademycouncil.net.

Intergovernmental Panel on Climate Change (IPCC) (1990), *Climate Change: The IPCC Scientific Assessment. Report Prepared for IPCC by Working Group I*, Cambridge, England: Cambridge University Press.

Intergovernmental Panel on Climate Change (IPCC) (1996), *Climate Change 1995: The Science of Climate Change. Contribution of Working Group I to the Second Assessment Report of the Intergovernmental Panel on Climate Change*, Cambridge, England: Cambridge University Press.

Intergovernmental Panel on Climate Change (IPCC) (2000), *Emissions Scenarios: A Special Report of Working Group III of the Intergovernmental Panel on Climate Change*, Cambridge, England: Cambridge University Press.

Intergovernmental Panel on Climate Change (IPCC) (2001), *Climate Change 2001: The Scientific Basis. Contribution of Working Group I to the Third Assessment Report of the Intergovernmental Panel on Climate Change*, Cambridge, England: Cambridge University Press.

Intergovernmental Panel on Climate Change (IPCC) (2005), *Notes for Lead Authors of the IPCC Fourth Assessment Report on Addressing Uncertainties*, Geneva: Intergovernmental Panel on Climate Change. Available at: http://www.ipcc.ch/pdf/supporting-material/uncertainty-guidance-note.pdf.

Intergovernmental Panel on Climate Change (IPCC) (2007a), *Climate Change 2007: The Physical Science Basis. Contribution of Working Group I to the Fourth Assessment Report of the Intergovernmental Panel on Climate Change*, Cambridge, England: Cambridge University Press.

Intergovernmental Panel on Climate Change (IPCC) (2007b), *Climate Change 2007: Impacts, Adaptation and Vulnerability. Contribution of Working Group II to the Fourth Assessment Report of the Intergovernmental Panel on Climate Change*, Cambridge, England: Cambridge University Press.

Intergovernmental Panel on Climate Change (IPCC) (2007c), *Climate Change 2007: Mitigation of Climate Change. Contribution of Working Group III to the Fourth Assessment Report of the Intergovernmental Panel on Climate Change*, Cambridge, England: Cambridge University Press.

Janssen, P.H.M., and Heuberger, P.S.C. (1995), Calibration of process-oriented models, *Ecological Modelling* 83: 55–66.

Janssen, P.H.M., Petersen, A.C., van der Sluijs, J.P., Risbey, J.S., and Ravetz, J.R. (2003), *RIVM/MNP Guidance for Uncertainty Assessment and Communication: Quickscan Hints and Actions List*, Bilthoven: Netherlands Environmental Assessment Agency (MNP), National Institute for Public Health and the Environment (RIVM). Available at: http://leidraad.pbl.nl.

Jasanoff, S. (1990), *The Fifth Branch: Science Advisers as Policymakers*, Cambridge, MA: Harvard University Press.

Jasanoff, S., and Wynne, B. (1998), Science and decisionmaking, in S. Rayner and E.L. Malone (eds.), *Human Choice and Climate Change, Vol. 1: The Societal Framework*, Columbus, OH: Batelle Press, pp. 1–87.

Kaufmann, W.J., III, and Smarr, L.L. (1993), *Supercomputing and the Transformation of Science*, New York: Scientific American Library.

Keller, E.F. (2003), Models, simulation, and computer experiments, in H. Radder (ed.), *The Philosophy of Scientific Experimentation*, Pittsburgh, PA: Pittsburgh University Press, pp. 198–215.

Kirschenmann, P.P. (1982), Some thoughts on the ideal of exactness in science and philosophy, in J. Agassi and R.S. Cohen (eds.), *Scientific Philosophy Today: Essays in Honor of Mario Bunge*, Dordrecht, the Netherlands: Reidel, pp. 85–98.

Kirschenmann, P.P. (1985), Neopositivism, marxism, and idealization: some comments on Professor Nowaks paper, *Studies in Soviet Thought* 30: 219–235.

Kirschenmann, P.P. (2001, June 29), Onzekerheid en risico in wetenschap, maatschappij en ethiek [Uncertainty and risk in science, society and ethics], Parting Address, Vrije Universiteit Amsterdam. Amsterdam: VU Boekhandel/Uitgeverij. [In Dutch]

Klinke, A., and Renn, O. (2002), A new approach to risk evaluation and management: risk-based, precaution-based, and discourse-based strategies, *Risk Analysis* 22: 1071–1094.

Kloprogge, P., van der Sluijs, J.P., and Petersen, A.C. (2011), A method for the analysis of assumptions in model-based environmental assessments, *Environmental Modelling and Software* 26: 289–301.

Knight, F.H. ([1921] 2002), *Risk, Uncertainty and Profit*, Washington, DC: Beard Books.

Knol, A.B., Petersen, A.C., van der Sluijs, J.P., and Lebret, E. (2009), Dealing with uncertainties in environmental burden of disease assessment, *Environmental Health* 8: 21 (13 pp).

Koperski, J. (1998), Models, confirmation, and chaos, *Philosophy of Science* 65: 624–648.

Küppers, G., and Lenhard, J. (2006), Simulation and a revolution in modelling style: from hierarchical to network-like integration, in J. Lenhard, G. Küppers, and T. Shinn (eds.), *Simulation: Pragmatic Constructions of Reality—Sociology of the Sciences*, Vol. 25, pp. 89–106.

Kwa, C.L. (1987), Representations of nature mediating between ecology and science policy: the case of the International Biological Programme, *Social Studies of Science* 17: 413–442.

Kwa, C.L. (2002), Romantic and baroque conceptions of complex wholes in the sciences, in J. Law and A. Mol (eds.), *Complexities: Social Studies of Knowledge Practices*, Durham, NC: Duke University Press, pp. 23–52.

Lahsen, M. (2008), Experiences of modernity in the greenhouse: a cultural analysis of a physicist 'trio' supporting the backlash against global warming, *Global Environmental Change* 18: 204–219.

Lakatos, I. (1970), Falsification and the methodology of scientific research programmes, in I. Lakatos and A. Musgrave (eds.), *Criticism and the Growth of Knowledge*, Cambridge, England: Cambridge University Press, pp. 91–196.

Laudan, L. (1984), *Science and Values: The Aims of Science and Their Role in Scientific Debate*, Berkeley: University of California Press.

Leggett, J. (1999), *The Carbon War: Dispatches from the End of the Oil Century*, London: Allen Lane.

Longino, H.E. (1990), *Science as Social Knowledge: Values and Objectivity in Scientific Inquiry*, Princeton, NJ: Princeton University Press.

Maas, R. (2007), Fine particles: from scientific uncertainty to policy strategy, *Journal of Toxicology and Environmental Health* A 70: 365–368.

Mann, M.E., Bradley, R.S., and Hughes, M.K. (1999), Northern Hemisphere temperatures during the past millennium: inferences, uncertainties, and limitations, *Geophysical Research Letters* 26: 759–762.

Markoff, J. (1994, November 24), Flaw undermines accuracy of Pentium chips, *New York Times*, p. D1.

Mastrandrea, M.D., Field, C.B., Stocker, T.F., Edenhofer, O., Ebi, K.L, Frame, D.J., Held, H., Kriegler, E., Mach, K.J., Matschoss, P.R., Plattner, G.-K., Yohe, G.W., and Zwiers, F.W. (2010), *Guidance Note for Lead Authors of the IPCC Fifth Assessment Report on Consistent Treatment of Uncertainties*, Geneva: Intergovernmental Panel on Climate Change.

Mayo, D.G. (1996), *Error and the Growth of Experimental Knowledge*, Chicago: University of Chicago Press.

McAvaney, B.J., Covey, C., Joussaume, S., Kattsov, V., Kitoh, A., Ogana, W., Pitman, A.J., Weaver, A.J., Wood, R.A., and Zhao, Z.-C. (2001), Model evaluation, in IPCC, *Climate Change 2001: The Scientific Basis. Contribution of Working Group I to the Third Assessment Report of the Intergovernmental Panel on Climate Change*, Cambridge, England: Cambridge University Press, pp. 471–523.

McGuffie, K., and Henderson-Sellers, A. (1997), *A Climate Modelling Primer*, 2nd edition, New York: Wiley.

McIntyre, S., and McKitrick, R. (2005), Hockey sticks, principal components, and spurious significance, *Geophysical Research Letters* 32: L03710, doi:10.1029/2004GL021750.

McMullin, E. (1985), Galilean idealization, *Studies in History and Philosophy of Science* 16: 247–273.

Merz, M. (1999), Multiplex and unfolding: computer simulation in particle physics, *Science in Context* 12: 293–316.

Merz, M., and Knorr Cetina, K. (1997), Deconstruction in a thinking science: theoretical physicists at work, *Social Studies of Science* 27: 73–111.

Meyer, L.A., and Petersen, A.C. (eds.) (2010), *Assessing an* IPCC *Assessment: An Analysis of Statements on Projected Regional Impacts in the 2007 Report*, The Hague: PBL Netherlands Environmental Assessment Agency.

Miller, C. (2001), Hybrid management: boundary organizations, science policy, and environmental governance in the climate regime, *Science, Technology, and Human Values* 26: 478–500.

Mitchell, J.F.B., Karoly, D.J., Hegerl, G.D., Zwiers, F.W., Allen, M.R., and Marengo, J. (2001), Detection of climate change and attribution of causes, in IPCC, *Climate Change 2001: The Scientific Basis. Contribution of Working Group I to the Third Assessment Report of the Intergovernmental Panel on Climate Change*, Cambridge, England: Cambridge University Press, pp. 695–738.

Morgan, M.S. (2003), Experiments without material intervention: model experiments, virtual experiments, and virtually experiments, in H. Radder (ed.), *The Philosophy of Scientific Experimentation*, Pittsburgh, PA: Pittsburgh University Press, pp. 216–235.

Morrison, M., and Morgan, M.S. (1999), Models as mediating instruments, in M.S. Morgan and M. Morrison (eds.), *Models as Mediators: Perspectives on Natural and Social Science*, Cambridge, England: Cambridge University Press, pp. 10–37.

Morton, A. (1993), Mathematical models: questions of trustworthiness, *British Journal for the Philosophy of Science* 44: 659–674.

Moss, R.H., and Schneider, S.H. (2000), Uncertainties in the IPCC TAR: recommendations to lead authors for more consistent assessment and reporting, in R. Pachauri, T. Taniguchi, and K. Tanaka (eds.), *Guidance Papers on the Cross Cutting Issues of the Third Assessment Report of the IPCC*, Geneva: Intergovernmental Panel on Climate Change, pp. 33–51.

Murphy, J.M., Sexton, D.M.H., Barnett, D.N., Jones, G.S., Webb, M.J., Collins, M., and Stainforth, D.A. (2004), Quantification of modelling uncertainties in a large ensemble of climate change simulation, *Nature* 430: 768–772.

National Institute for Public Health and the Environment/Netherlands Environmental Assessment Agency (RIVM/MNP) (2003), *Nuchter omgaan met risicos [Coping Rationally with Risks]*, Report 251701047, Bilthoven: Netherlands Environmental Assessment Agency (MNP), National Institute for Public Health and the Environment (RIVM). Available at: http://www.rivm.nl/bibliotheek/rapporten/251701047.pdf. [In Dutch.]

National Research Council ([1994] 1996a), *Science and Judgment in Risk Assessment*, Washington, DC: Taylor & Francis.

National Research Council (1996b), *Understanding Risk: Informing Decisions in a Democratic Society*, Washington, DC: National Academy Press.

National Research Council (1998), *The Capacity of U.S. Climate Modeling to Support Climate Change Assessment Activities*, Washington, DC: National Academy Press.

Naylor, T.H. (ed.) (1966), *Computer Simulation Techniques*, New York: Wiley.

Nebeker, F. (1995), *Calculating the Weather: Meteorology in the 20th Century*, San Diego, CA: Academic Press.

Nickels, T. (1988), Reconstructing science: discovery and experiment, in D. Batens and J.P van Bendegem (eds.), *Theory and Experiment*, Dordrecht, the Netherlands: Reidel, pp. 33–53.

Norton, S.D., and Suppe, F. (2001), Why atmospheric modeling is good science, in C.A. Miller and P.N. Edwards (eds.), *Changing the Atmosphere: Expert Knowledge and Environmental Governance*, Cambridge, MA: MIT Press, pp. 67–105.

Nowak, L. (1985), Marxism and positivism, or, dialectics in books and dialectics in action, *Studies in Soviet Thought* 30: 195–218.

Oreskes, N., Shrader-Frechette, K., and Belitz, K. (1994), Verification, validation, and confirmation of numerical models in the earth sciences, *Science* 263: 641–646.

Palmer, T. (2005), Global warming in a nonlinear climate: can we be sure?, *Europhysics News* 36: 42–46.

Parker, W.S. (2006), Understanding pluralism in climate modeling, *Foundations of Science* 11: 349–368.

Parker, W.S. (2009), Confirmation and adequacy-for-purpose in climate modeling, *Proceedings of the Aristotelian Society* 83 (Suppl.): 233–249.

Petersen, A.C. (1995), Consensus in de natuurwetenschappen: de rationaliteit van consensus in het klimaatonderzoek [Consensus in the natural sciences: The rationality of consensus in climate research], M.A. thesis, Faculty of Philosophy, Vrije Universiteit, Amsterdam, IMAU Report 95-21. [In Dutch]

Petersen, A.C. (1999a, October 21), Clinton maakte het senaat te makkelijk [Clinton made things too easy for the senate], *de Volkskrant*. [In Dutch]

Petersen, A.C. (1999b), Convection and chemistry in the atmospheric boundary layer, PhD dissertation, Utrecht, the Netherlands: Utrecht University.

Petersen, A.C. (2000a), The impact of chemistry on flux estimates in the convective boundary layer, *Journal of the Atmospheric Sciences* 57: 3398–3405.

Petersen, A.C. (2000b), Models as technological artefacts, *Social Studies of Science* 30: 793–799.

Petersen A.C. (2000c), Philosophy of climate science, *Bulletin of the American Meteorological Society* 81: 265–271.

Petersen, A.C. (2004), Models and geophysiological hypotheses, in S.H. Schneider, J.R. Miller, E. Crist, and P.J. Boston (eds.), *Scientists Debate Gaia: The Next Century*, Cambridge, MA: MIT Press, pp. 37–44.

Petersen, A.C., Beets, C., van Dop, H., Duynkerke, P.G., and Siebesma, A.P. (1999), Mass-flux characteristics of reactive scalars in the convective boundary layer, *Journal of the Atmospheric Sciences* 56: 37–56.

Petersen, A.C., Cath, A., Hage, M., Kunseler, E., and van der Sluijs, J.P. (2011), Post-normal science in practice at the Netherlands Environmental Assessment Agency, *Science, Technology, and Human Values* 36: 362–388.

Petersen, A.C., and Holtslag, A.A.M. (1999), A first-order closure for covariances and fluxes of reactive species in the convective boundary layer, *Journal of Applied Meteorology* 38: 1758–1776.

Petersen, A.C., Janssen, P.H.M., van der Sluijs, J.P., Risbey, J.S., and Ravetz, J.R. (2003), *RIVM/MNP Guidance for Uncertainty Assessment and Communication: Mini-Checklist and Quickscan Questionnaire*, Bilthoven: Netherlands Environmental Assessment Agency (MNP), National Institute for Public Health and the Environment (RIVM). Available at: http://leidraad.pbl.nl.

Pickering, A. (ed.) (1992), *Science as Practice and Culture*, Chicago: University of Chicago Press.

Pielke, R.A. (2007), *The Honest Broker: Making Sense of Science in Policy and Politics*, Cambridge, England: Cambridge University Press.

Polanyi, M. (1962 [1958]), *Personal Knowledge: Towards a Post-Critical Philosophy*, London: Routledge.

Popper, K.R. ([1934] 1959), *The Logic of Scientific Discovery*, London: Routledge.

Popper, K.R. ([1972] 1979), *Objective Knowledge: An Evolutionary Approach*, Oxford, England: Clarendon Press.

Port, O., and Tashiro, H. (2004, June 7). Supercomputing, Asian cover story, *Business Week* 3886: 50–54.

Price, D.K. (1965), *The Scientific Estate*, Cambridge, MA: Harvard University Press.

Proctor, R.N. (1991), *Value-Free Science: Purity and Power in Modern Knowledge*, Cambridge, MA: Harvard University Press.

Radder, H. (1982), Between Bohrs atomic theory and Heisenbergs matrix mechanics: a study of the role of the Dutch physicist H.A. Kramers, *Janus* 69: 223–252.

Radder, H. ([1984] 1988), *The Material Realization of Science: A Philosophical View on the Experimental Natural Sciences, Developed in Discussion with Habermas*, Assen, the Netherlands: Van Gorcum.

Radder, H. (1996), *In and About the World: Philosophical Studies of Science and Technology*, Albany: State University of New York Press.

Randall, D.A., and Wielicki, B.A. (1997), Measurements, models, and hypotheses in the atmospheric sciences, *Bulletin of the American Meteorological Society* 78: 399–406.

Ravetz, J.R. (1996 [1971]), *Scientific Knowledge and Its Social Problems*, New Brunswick, NJ: Transaction.

Rayner, S., and Malone, E.L. (eds.) (1998), *Human Choice and Climate Change* (4 vols.), Columbus, OH: Batelle Press.

Redhead, M. (1980), Models in physics, *British Journal for the Philosophy of Science* 31: 145–163.

Rind, D. (1999), Complexity and climate, *Science* 284: 105–107.

Risbey, J., van der Sluijs, J., Kloprogge, P, Ravetz, J., Funtowicz, S., and Corral Quintana, S. (2005), Application of a checklist for quality assistance in environmental modelling to an energy model, *Environmental Modeling and Assessment* 10: 63–79.

Rohrlich, F. (1991), Computer simulation in the physical sciences, in A. Fine, M. Forbes, and L. Wessels (eds.), *PSA 1990, Vol. 2: Symposium and Invited Papers*, East Lansing, MI: Philosophy of Science Association, pp. 497–506.

Rotmans, J., and de Vries, H.J.M. (eds.) (1997), *Perspectives on Global Change: The TARGETS Approach*, Cambridge, England: Cambridge University Press.

Rotmans, J., and Dowlatabadi, H. (1998), Integrated assessment modeling, in S. Rayner and E.L. Malone (eds.), *Human Choice and Climate Change, Vol. 3: Tools for Policy Analyis*, Columbus, OH: Batelle Press, pp. 291–377.

Rouse, J. (1987), *Knowledge and Power: Toward a Political Philosophy of Science*, Ithaca, NY: Cornell University Press.

Saloranta, T.M. (2001), Post-normal science and the global climate change issue, *Climatic Change* 50: 395–404.

Saltelli, A. (2000), What is sensitivity analysis? in A. Saltelli, K. Chan and E.M. Scott (eds.), *Sensitivity Analysis*, Chichester, England: John Wiley & Sons, pp. 3–13.

Scheffer, M., Brovkin, V., and Cox P.M. (2006), Positive feedback between global warming and atmospheric CO_2 concentration inferred from past climate change, *Geophysical Research Letters* 33: L10702, doi:10.1029/2005GL025044, 4 pp.

Shackley, S. (2001), Epistemic lifestyles in climate change modeling, in C.A. Miller and P.N. Edwards (eds.), *Changing the Atmosphere: Expert Knowledge and Environmental Governance*, Cambridge, MA: MIT Press, pp. 107–133.

Shackley, S., Risbey, J., Stone, P., and Wynne, B. (1999), Adjusting to policy expectations in climate change modeling: an interdisciplinary study of flux adjustments in coupled atmosphere-ocean general circulation models, *Climatic Change* 43: 413–454.

Shackley, S., and Skodvin, T. (1995), IPCC gazing and the interpretative social sciences: a comment on Sonja Boehmer-Christiansen's 'Global climate policy: The limits of scientific advice', *Global Environmental Change* 5: 175–180.

Shackley, S., and Wynne, B. (1996), Representing uncertainty in global climate change science and policy: boundary-ordering devices and authority *Science, Technology, and Human Values* 21: 275–302.

Shackley, S., Young, P., Parkinson, S., and Wynne, B. (1998), Uncertainty, complexity and concepts of good science in climate change modeling: are GCMs the best tools? *Climatic Change* 38: 155–201.

Siekmann, H. (1998), Experiment und Computersimulation in der Strömungstechnik, in M. Heidelberger and F. Steinle (eds.), *Experimental Essays—Versuche zum Experiment*, Baden-Baden, Germany: Nomos Verlagsgesellschaft, pp. 209–226.

Sismondo, S. (1999), Models, simulations, and their objects, *Science in Context* 12: 247–260.

Skodvin, T. (2000), *Structure and Agent in the Scientific Diplomacy of Climate Change: An Empirical Case Study of Science-Policy Interaction in the Intergovernmental Panel on Climate Change*, Dordrecht, the Netherlands: Kluwer Academic.

Slovic, P., Finucane, M.L., Peters, E., and MacGregor, D.G. (2004), Risk as analysis and risk as feelings: some thoughts about affect, reason, risk, and rationality, *Risk Analysis* 24: 311–322.

Smith, L.A. (2002), What might we learn from climate forecasts?, *Proceedings of the National Academy of Sciences of the United States of America* 99 (Suppl. 1): 2487–2492.

Social Learning Group (2001), *Learning to Manage Global Environmental Risks, Volume 1: A Comparative History of Social Responses to Climate Change, Ozone Depletion, and Acid Rain*, Cambridge, MA: MIT Press.

Star, S.L., and Griesemer, J.R. (1989), Institutional ecology, 'translations', and boundary objects: amateurs and professionals in Berkeley's Museum of Vertebrate Zoology, 1907–39, *Social Studies of Science* 19: 387–420.

Stevens, B., and Lenschow, D.H. (2001), Observations, experiments, and large eddy simulation, *Bulletin of the American Meteorological Society* 82: 283–294.

Swart, R., Bernstein, L., Ha-Duong, M., and Petersen, A.C. (2009), Agreeing to disagree: uncertainty management in assessing climate change, impacts and responses by the IPCC, *Climatic Change* 92: 1–29.

Tennekes, H. (1994), The limits of science, in W. Zweers and J.J. Boersema (eds.), *Ecology, Technology and Culture*, Cambridge, England: White Horse Press, pp. 72–88.

Thompson, M., Ellis, R., and Wildavsky, A. (1990), *Cultural Theory*, Boulder, CO: Westview Press.

Torn, M.S., and Harte, J. (2006), Missing feedbacks, asymmetric uncertainties, and the underestimation of future warming, *Geophysical Research Letters* 33: L10703, doi:10.1029/2005GL025540 (5 pp).

United Nations (1992), *United Nations Framework Convention on Climate Change*. Available at: http://www.unfccc.int.

United Nations Educational, Scientific and Cultural Organization (UNESCO) (2005), *The Precautionary Principle*, Report of the World Commission on the Ethics of Scientific Knowledge and Technology (COMEST), Paris: UNESCO.

van Asselt, H., Berghuis, J., Biermann, F., Cornelisse, C., Haug, C., Gupta, J., and Massey, E. (2008), *Exploring the Socio-Political Dimensions of Climate Change Mitigation*, IVM Report W-08/18, Amsterdam: Institute for Environmental Studies, VU University Amsterdam. Available at: http://www.ivm.vu.nl/images_upload/4BC45814-D58C-1894-9C415605D27FB5B1.pdf.

van Asselt, M.B.A. (2000), *Perspectives on Uncertainty and Risk: The PRIMA Approach to Decision Support*, Dordrecht, the Netherlands: Kluwer Academic.

van Asselt, M.B.A., and Petersen, A.C. (eds.) (2003), *Niet bang voor onzekerheid* [*Not Afraid of Uncertainty*], Utrecht, the Netherlands: Lemma Publishers. [In Dutch. Published for the Netherlands Advisory Council for Research on Spatial Planning, Nature and the Environment (RMNO).]

van Asselt, M.B.A., and Rotmans, J. (1997), Uncertainties in perspective, in J. Rotmans and H.J.M. de Vries (eds.), *Perspectives on Global Change: The TARGETS Approach*, Cambridge, England: Cambridge University Press, pp. 207–222.

van der Sluijs, J.P. (1995), Uncertainty management in integrated modelling: the IMAGE case, in S. Zwerver, R.S.A.R. van Rompaey, M.T.J. Kok, and M.M. Berk (eds.), *Climate Change Research: Evaluation and Policy Implications*, Amsterdam: Elsevier Science, pp. 1401–1406.

van der Sluijs, J.P. (1997), Anchoring amid uncertainty: On the management of uncertainties in risk assessment of anthropogenic climate change, Ph.D. dissertation, Utrecht, the Netherlands: Utrecht University.

van der Sluijs, J.P. (2002), A way out of the credibility crisis of models used in integrated environmental assessment, *Futures* 34: 133–146.

van der Sluijs, J.P., Craye, M., Funtowicz, S., Kloprogge, P., Ravetz, J., and Risbey, J. (2005), Combining quantitative and qualitative measures of uncertainty in model based environmental assessment: the NUSAP system, *Risk Analysis* 25: 481–492.

van der Sluijs, J.P., Janssen, P.H.M., Petersen, A.C., Kloprogge, P., Risbey, J.S., Tuinstra, W., and Ravetz, J.R. (2004), *RIVM/MNP Guidance for Uncertainty Assessment and Communication: Tool Catalogue for Uncertainty Assessment*, Utrecht, the Netherlands: Utrecht University. Available at: http://www.nusap.net/downloads/toolcatalogue.pdf.

van der Sluijs, J.P., Risbey, J.S., Kloprogge, P., Ravetz, J.R., Funtowicz, S.O., Coral Quintana, S., Pereira, A.G., De Marchi, B., Petersen, A.C., Janssen, P.H.M., Hoppe, R., and Huijs, S.W.F. (2003), *RIVM/MNP Guidance for Uncertainty Assessment and Communication: Detailed Guidance*, Utrecht, the Netherlands: Utrecht University. Available at: http://www.nusap.net/downloads/detailedguidance.pdf.

van der Sluijs, J.P., van Eijndhoven, J.C.M., Wynne, B., and Shackley, S. (1998), Anchoring devices in science for policy: the case of consensus around climate sensitivity, *Social Studies of Science* 28: 291–323.

van der Sluijs, J.P., Petersen, A.C., Janssen, P.H.M., Risbey, J.S., and Ravetz, J.R. (2008), Exploring the quality of evidence for complex and contested policy decisions, *Environmental Research Letters* 3: 024008 (9 pp).

van Egmond, N.D. (1999, February 3), Modellen geven meetresultaten betekenis [Models give meaning to measurement results], *Trouw*, p. 16. [In Dutch]

Vasileiadou, E., Heimeriks, G., and Petersen, A.C. (2011), Exploring the impact of the IPCC Assessment Reports on science, *Environmental Science & Policy* 11: 627–641.

Visser, H., Büntgen, U., D'Arrigo, R., and Petersen, A.C. (2010), Detecting instabilities in tree-ring proxy calibration, *Climate of the Past* 6: 367–377.

von Storch, H. (2001), Models between academia and applications, in H. von Stoch and G. Flöser (eds.), *Models in Environmental Research*, Berlin: Springer-Verlag, pp. 17–33.

Walker, W.E., Harremoës, P., Rotmans, J., van der Sluijs, J.P., van Asselt, M.B.A., Janssen, P.H.M., and Krayer von Krauss, M.P. (2003), Defining uncertainty: a conceptual basis for uncertainty management in model-based decision support, *Integrated Assessment* 4: 5–17.

Wardekker, J.A., van der Sluijs, J.P., Janssen, P.H.M., Kloprogge, P., and Petersen, A.C. (2008), Uncertainty communication in environmental assessments: views from the Dutch science–policy interface, *Environmental Science and Policy* 11: 627–641.

Weinberg, A.M. (1972), Science and trans-science, *Minerva* 10: 209–222.

Weinert, F. (1999), Theories, models and constraints, *Studies in History and Philosophy of Science* 30: 303–333.

Werner, B.T. (1999), Complexity in natural landform patterns, *Science* 284: 102–104.

Wiener, J.B., and Rogers, M.D. (2002), Comparing precaution in the United States and Europe, *Journal of Risk Research* 5: 317–349.

Winsberg, E. (2003), Simulated experiments: methodology for a virtual world, *Philosophy of Science* 70: 105–125.

Winsberg, E. (2010), *Science in the Age of Computer Simulation*, Chicago: University of Chicago Press.

Wynne, B. (1992), Uncertainty and environmental learning: reconceiving science and policy in the preventive paradigm, *Global Environmental Change* 6: 111–127.

Appendix: Proceedings and Discussion of the IPCC Contact Group Meeting on Attribution, 20 January 2001, Shanghai

In this appendix, I present the proceedings (indicated as block quotations) and my discussion of the contact group meeting on the final paragraph of the Intergovernmental Panel on Climate Change (IPCC) (2001) detection and attribution section in the Summary for Policymakers that was held in Shanghai on 20 January 2001.* I was present at this meeting as a philosophical observer within the Dutch delegation. I sat at the table next to the lead authors in the small meeting room but did not participate in the discussion.

The session started at 8:00 a.m. and was chaired by the chair of the IPCC. Here is the text of the paragraph that was distributed to the delegates before the beginning of the plenary session (the Shanghai Draft) (changes with respect to the Final Draft are marked by striking through and underlining):

> ~~It is likely that increasing concentrations of anthropogenic greenhouse gases have contributed substantially to the observed warming over the last 50 years. Nevertheless, the accuracy~~The precision ~~accuracy~~of estimates of ~~the magnitude of anthropogenic warming, and particularly of~~ the contribution from ~~influence of the~~ individual ~~external factors,~~factors to recent climate change continues to be limited by uncertainties in ~~estimates of~~ internal variability, natural and anthropogenic forcing ~~radiative factors~~, in particular that ~~the forcing~~by anthropogenic aerosols, and the estimated climate response ~~to those factors~~. _Despite these uncertainties, it is likely that increasing concentrations of anthropogenic greenhouse gases have contributed substantially to the observed warming over the last 50 years._

This was the standard way in which revisions of text were presented in Shanghai. The Shanghai Draft itself had also been distributed in a striked-through/underlined version.†

> The Chair opens the contact group session and sums up three questions that need to be addressed: (1) the language of the paragraph (e.g., the use of the word 'substantially'), (2) the ordering of the sentences, and (3) the way uncertainties are characterised. He proposes to discuss the issue of ordering at the end of the session. He then goes through the section text

---

* I have sent my transcript of the proceedings (edited on the basis of my notes) to one of the coordinating lead authors of the detection and attribution chapter to have him check its accuracy and have received a positive response.

† See the box in Chapter 7 for more readable versions of both the Final Draft and the Shanghai Draft.

that has already been approved and concludes his introduction by noting that there was much support in the plenary session for keeping the word 'substantially' in the text and that the lead authors had worked on the precise wording for one and a half years.

The main objection initially raised by one country (country B*) during the plenary was that the word *substantial* could not adequately be translated into its own language, one of the U.N. languages.[†] Usually, such arguments are used for lack of real arguments. It is taken to signal that some country just wants to get rid of a specific word for political reasons. Most people who were present in Shanghai thought that this was the case for country B: They just wanted to get rid of 'substantial' to have it replaced, preferably by something weaker (e.g., 'discernible').

The Language of the Final Sentence of the Draft

Country B says it has problems with the word 'substantial', since it is not a scientifically defined term. B states that they would like to see numbers in the text. In response to the Chair's question of which wording they would like to have as a substitute, B provides an alternative: 'comparable with, or larger than'. The Chair verifies with B that they think that the use of phrases such as 'largely due to' or 'most of' would do instead of 'substantial'. B replies that they have similar problems with 'largely due to', since also that is a subjective phrase.

At this point, many of the participants were caught by surprise. Instead of proposing to weaken *substantial* to something like *discernible* (as was used five years earlier in the Second Assessment Report [SAR]), country B asked for a stronger and quantified claim. From a speculative point of view (circumstantial evidence follows during discussion of the proceedings), I think the reason why country B wanted to push the lead authors toward a quantified comparative statement is to better expose the epistemological issue that, in this case, *models* are compared to observations. By using the word *substantial*, the fact that a model simulation is used becomes less clear to the reader. Country B thinks that if, in this conclusion, the strongest model statement of the whole detection and attribution section (that the greenhouse gas signal of the last 50 years was estimated to be *comparable with, or larger than* the

---

* To focus on the substance of what was said, here the seven countries that played an identifiable role in this contact group meeting are labelled B through J (where B stands for Saudi Arabia; see Chapter 7, page 161, first note).

† A second country (France), within a different political group, supported Saudi Arabia's intervention by stating *substantial* was also difficult to translate in French (like Arabic, a U.N. language).

observed warming) is emphasised, it will subsequently be easier to dismiss the conclusion as 'just based on models' (since country B prefers to hold and proclaim the view that all climate simulations are unreliable). In particular, the larger than phrase is attractive for this purpose, as discussed further in this appendix.

The reason why the lead authors had come up with a relatively weak conclusion at the end of the detection and attribution section was that they wanted not only to give an estimate of the inexactness of the claim (in terms of estimated model error) but also to take unreliability$_2$ into account (by using the word *likely* instead of *very likely*). The lead authors are at this point not judging the stronger modelling statement to be reliable$_2$ enough to be 'likely' true.

> The Chair asks B how they would like the entire sentence to read. B replies that they would also like to see a reference to the use of models in the sentence. The lead author who is operating the laptop that is used to project alternative sentences on the screen objects that it is the observations that lead to this conclusion. Furthermore, one of the co-ordinating lead authors objects to changing the text from 'substantial' to 'comparable with, or larger than', since the current statement containing 'substantial' is a rather weak one. The co-ordinating lead author tells B that he does not understand why B thinks the sentence would be mis-interpreted. B replies that attribution involves a comparison of models with observations.

The laptop lead author gets a sense of what country B is up to, but he over-reacts. It is obvious that model results play an important role (see Chapter 5). What the lead author wants to say here, granting him a charitable reading, is that particular types of uncertainty (e.g., that due to uncertainty in climate sensitivity) can be taken into account in detection and attribution studies (IPCC 2001, SPM: 10).* The coordinating lead author, however, is very frank about the fact that it is a considered judgement that is presented here.

> Country C (represented by a review editor of the detection and attribu-tion chapter) proposes an alternative wording to accommodate some of the concerns of B: 'The warming attributable to increasing concentrations of anthropogenic greenhouse gases is likely to be comparable with the observed warming over the last 50 years'. After this proposal the lead authors discuss among themselves whether or not this statement (using the word 'comparable') is scientifically justified, or not (i.e., too strong).

Here, some progress is made towards a solution. Country C's delegate is very influential within the IPCC, being a member of the IPCC Working Group I (WG I) Bureau. People listen to him. He happened to be the review

---

* In references to IPCC reports, aside from the page number in the whole report, the part of the report is also included: for example, SPM = Summary for Policymakers; TS = Technical Summary; Ch. *x* = Chapter *x*; or Glossary.

editor for the relevant chapter, so he knows exactly what everyone's motive is. The sentence he crafts here completely matches the expectations of country B, but it turns out to be still a bit too strong for the lead authors, as I discuss further in this appendix.

> Country D asks whether the authors only would like to show that there is a correlation, without specifying the magnitude. In that case 'substantial' would be just as good as 'comparable'.

It was unclear, I imagine, to most participants what could possibly be meant by using the word *comparable* without making a quantitative comparison—and thus specifying the magnitude. The intervention was ignored, anyhow.

> The Chair remarks that the phrase 'comparable with' has already been used in an earlier bullet point in the text ('Most of these studies find that, over the last 50 years, the estimated rate and magnitude of warming due to increasing concentrations of greenhouse gases alone are comparable with, or larger than, the observed warming'.), that it is not a matter of comparing model results to observations, and that the contact group should come up with a real conclusion for the section.

Here, we see that the new sentence that was added to the SPM in Shanghai is used by the chair to justify a rejection of country B's request. But, the chair's observation is not new to country B. In their proposal to use 'comparable with, or larger than', they explicitly referred to the earlier bullet. Country B also knows that the chair's statement 'that it is not a matter of comparing model results to observations' is incorrect. The chair is apparently slightly misled by the earlier intervention of the laptop lead author.

> The lead authors bring back the results of their internal discussion to the group. They repeat that since 'substantially' is weaker than 'largely due to', they would like to stick with the former phrase.

Country C's delegate had already suggested in the plenary that 'largely due to' be used. The lead authors now narrow down the options to either substantially or largely due to, while expressing a preference for the former phrase. Comparable with is not mentioned by the lead authors as an option that can be considered.

> Country E proposes to use the phrase 'first factor', to which the lead authors reply that there are problems with quantifying the other signals than the anthropogenic greenhouse-gas signal. Country F does not agree with the lead authors' reply and proposes to use 'principally' instead of 'substantially'. B supports E's intervention. Country G makes another proposal: 'dominant factor'. D: 'primary factor'. Chair: 'principal cause'. B states that they could agree with 'larger than'. The lead authors explain that 'substantial' was used in a non-quantified relative sense.

At this point in the meeting, it becomes clear that many countries want to see a stronger statement than the one containing 'substantial'. Country B is trying to make the most of it by changing their proposal to use comparable with, or larger than into larger than.

> E wants to change the word 'warming' in the final sentence to 'warming of the climate system'.

This intervention was ignored: it was not considered important at this point.

> C makes a new proposal: 'The observed warming over the last 50 years is *mainly due* to increasing concentrations of anthropogenic greenhouse gases' (thus reversing the order of the sentence). According to C, this makes the sentence flow better. This sentence is typed into the computer and appears on the screen. [To now, the lead author has acted quite chaotically in typing in alternatives (or refusing to do so) and scrolling sentences on and off the screen; the chair had to call him to order for the session to be able to proceed in a structured manner.] A subsequent intervention proposes to change the sentence to '*Most* of the observed warming over the last 50 years is *due* to increasing concentrations of anthropogenic greenhouse gases'. Another intervention proposes to delete 'anthropogenic'. E wants to have the phrase 'principal cause' in the sentence. [The two alternatives are simultaneously displayed on the screen—with the 'principal cause' formulation on top.] C suggests to insert 'now likely' before 'due to'. This is not typed onto the screen, however.

Here, country C's delegate makes the proposal that will more or less finally prevail. 'Mainly due to' is inconspicuously changed into 'most'. One word, however, is still missing—despite the request to put it on the screen: *likely*.

> The lead authors could accept the 'principal cause' formulation, but repeat that they really prefer to keep the original formulation using 'substantially'. [Although it becomes clear that the lead authors disagree among themselves at this stage, it is especially the lead author at the laptop who objects to using 'most'.] B reminds the session that it should be an IPCC WG I judgment (not only a judgment by the lead authors). B makes clear it could go with the second formulation on the screen [the one containing 'most'] and not with formulations using 'principal cause' or 'substantially'. Country H objects to the 'principal cause' formulation: other causes should be mentioned here as well, otherwise the sentence would be unbalanced. The Chair reminds H that the other causes are already mentioned in other paragraphs of the section. C says that 'most' is good. G thinks that the use of 'most' will naturally lead to the question 'what about the rest?' and is therefore not entirely happy with the word. Country J reminds the session to insert 'It is likely that' before 'most' in the sentence.

The uncertainty phrase *likely* is not forgotten. *Most* appears to be on the winning side.

> The lead author at the laptop thinks that using 'most' would mean that one study showing a 45% contribution can refute the IPCC conclusion. J replies that this is not the case, since the word 'likely' is also used in the sentence. The lead authors first say that they had agreed that if 'substantial' was not possible, they would be willing to go along with C's earlier proposal to use 'largely due to'. It finally turns out that they are now also willing to agree with 'most'.

Most is clearly pushing the lead authors to the limit of what they are willing to defend. But now that the lead authors have accepted 'most', the race is run, as will become obvious next.

> E does not want to start the sentence with 'It is likely that'. The lead authors [especially the one at the laptop] agree and immediately delete the phrase from the screen. The Chair gets angry with the lead author at the laptop, forces him to reinsert 'It is likely that' and tells him that this is an intergovernmental meeting and the lead authors are only allowed to say whether something is wrong or not.

This was a clear example of a power play by the chair. Apart from structuring procedures, leadership plays a central role in the IPCC (Skodvin 2000).

> G proposes to remove 'most'. The lead authors do not agree now. Also E is against 'most'. The Chair decides to close the discussion on the language of the sentence by proposing to keep 'most' [he puts some pressure on the delegates by saying that the session has to move on since time is running out]. No one objects.

Thus there was no longer a possibility for countries to get rid of 'most' after the lead authors had agreed.

Characterisation of Uncertainties and Ordering

> C proposes to discuss the uncertainties in the same sentence using the qualifier: 'In light of the new evidence and taking into account remaining uncertainties, ...' D does not want to have a qualifier at the beginning of the sentence. B, however, does. The Chair suggests that one could do without the qualifier. C agrees that it can be removed again. H wants to put 'in light of the new evidence' later in the sentence and make a second sentence discussing the uncertainties. D repeats that it wants to get rid of the qualifier. The Chair proposes to keep the sentence as it now is (i.e., including the qualifier). E does not agree and wants to have the qualifier removed. The Chair asks the countries to give their vote. There appears to be a split among the

countries. The Chair asks the participants to accept the sentence as it appears now on the screen as a compromise [several minor editorial changes were made, not documented as interventions in the preceding material]: 'In the light of the new evidence and taking into account the remaining uncertainties, most of the observed warming over the last 50 years is likely to have been due to the increase in greenhouse gas concentrations.' D is willing to accept the sentence as a compromise. Also E agrees. No one else objects. The session is closed.

Now with the session being closed, the word *most* trickles down to the title of the section as well, which—after extensive discussion not documented here—becomes

> There is new and stronger evidence that most of the warming observed over the last 50 years is attributable to human activities. (IPCC 2001, SPM: 10)

This is the message that is sent out to the world. In the media, it has become 'it is certain that most ...'. And country B used the first opportunity in the climate convention's Subsidiary Body for Scientific and Technological Advice (SBSTA) where the IPCC Third Assessment Report (TAR) was discussed to air its view that the science 'is still very uncertain'.

Index